Lecture Notes in Computer Science 8296

Jin Akiyama Mikio Kano Toshinori Sakai (Eds.)

Computational Geometry and Graphs

Thailand-Japan Joint Conference, TJJCCGG 2012
Bangkok, Thailand, December 6-8, 2012
Revised Selected Papers

Volume Editors

Jin Akiyama
Tokyo University of Science
Research Center for Math and Science Education
Kagurazaki, Shinjuku-ku, Tokyo, Japan
E-mail: ja@jin-akiyama.com

Mikio Kano
Ibaraki University
Department of Computer and Information Science
Hitachi, Ibaraki, Japan
E-mail: kano@mx.ibaraki.ac.jp

Toshinori Sakai
Tokai University
Research Institute of Educational Development
Tomigaya, Shibuya-ku, Tokyo, Japan
E-mail: sakai@tokai-u.jp

ISSN 0302-9743 e-ISSN 1611-3349
ISBN 978-3-642-45280-2 e-ISBN 978-3-642-45281-9
DOI 10.1007/978-3-642-45281-9
Springer Heidelberg New York Dordrecht London

Library of Congress Control Number: 2013954768

CR Subject Classification (1998): G.2, I.3.5, F.2

LNCS Sublibrary: SL 1 – Theoretical Computer Science and General Issues

Typesetting: Camera-ready by author, data conversion by Scientific Publishing Services, Chennai, India

Printed on acid-free paper

Springer is part of Springer Science+Business Media (www.springer.com)

Preface

The Thailand-Japan Joint Conference on Computational Geometry and Graphs (TJJCCGG 2012) was held during December 6–8, 2012, at Srinakharinwirot University, Bangkok, Thailand. Previous conferences were held in Tokyo as JCDCG 1997, 1998, 2000, 2002, 2004 (Japan Conference on Discrete and Computational Geometry), in Kyoto as KyotoCGGT 2007 (Kyoto International Conference on Computational Geometry and Graph Theory), and in Kanazawa as JCCGG 2009 (Japan Conference on Computational Geometry and Graphs). Other conferences in this series were also held in Manila (2001), Bandung (2003), Tianjin (2005), and Dalian (CGGA 2010).

TJJCCGG 2012 provided a forum for researchers working in computational geometry, graph theory/algorithms, and their applications. This proceedings volume consists of original research papers in these areas. Applied and experimental papers in this volume show convincingly the usefulness and efficiency of algorithms in a practical setting.

This volume contains 15 papers selected from among six plenary talks, one special public talk, and 41 talks by participants from about 20 countries around the world. The papers have been carefully peer-reviewed by experts and revised before acceptance.

This conference was dedicated to Prof. Narong Punnim and Prof. Wanida Hemakul for their significant contributions to Thai mathematics.

September 2013

Jin Akiyama
Mikio Kano
Toshinori Sakai

Organization

Conference Co-chairs

Jin Akiyama	Tokai University, Japan
Narong Punnim	Srinakharinwirot University, Thailand

Program Co-chairs

Mikio Kano	Ibaraki University, Japan
Wanida Hemakul	Chulalongkorn University, Thailand
Toshinori Sakai	Tokai University, Japan

Organizing Co-chairs

Nittiya Pabhapote	University of the Thai Chamber of Commerce, Thailand
Sermsri Thaithae	Srinakharinwirot University, Thailand
Sayun Sotaro	Srinakharinwirot University, Thailand
Varaporn Seanpholphat	Srinakharinwirot University, Thailand
Chariya Uiyyasathian	Chulalongkorn University, Thailand
Ruangvarin Saranraksakul	Srinakharinwirot University, Thailand
Deaw Jaibun	Mahidol Wittayanusorn School, Thailand
Raweewon Ngamsuntikul	Srinakharinwirot University, Thailand

Program Committee

Mikio Kano	Ibaraki University, Japan
Hiro Ito	Kyoto University, Japan
Toshinori Sakai	Tokai University, Japan
Stefan Langerman	Université Libre de Bruxelles, Belgium
G.L. Chia	University of Malaya, Malaysia
Wanida Hemakul	Chulalongkorn University, Thailand

Table of Contents

Operators which Preserve Reversibility

Jin Akiyama[1] and Hyunwoo Seong[2]

[1] Research Center for Math Education, Tokyo University of Science,
1-3 Kagurazaka, Shinjuku, Tokyo 162-8601, Japan
ja@jin-akiyama.com

[2] Department of Mathematics, The University of Tokyo,
3-8-1 Komaba, Meguro, Tokyo 153-8914, Japan
hwseong@hotmail.com

Dedicated to Dr. Narong Punnim and Dr. Wanida Hemakul.

Abstract. A given pair of convex polygons α and β is said to be reversible if α and β have dissections into a common finite number of pieces which can be rearranged to form β and α respectively, under certain conditions. A polygon α is said to be reversible if there exists a polygon β such that the pair α and β is reversible. This paper discusses operators which preserve reversibility for polygons. All reversible polygons are classified into seven equivalence classes $\mathfrak{P}_i$ $(i = 1, 2, \ldots, 7)$ under the equivalence relation $\equiv$, where $A \equiv B$ means that there exists some operator f such that $B = f(A)$.

1 Introduction

In the beginning of the 20th century, Henry E. Dudeney ([1]) proposed the 'Haberdasher's Problem' of dissecting a regular triangle into a finite number of pieces which can be rearranged to form a square (Fig. 1).

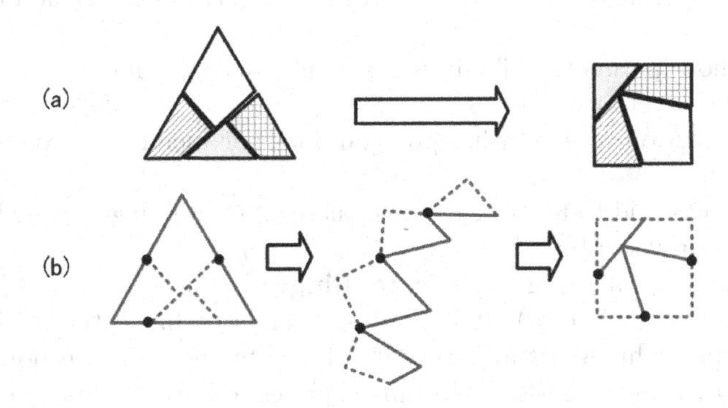

Fig. 1. (a) The answer to Dudeney's haberdasher's puzzle (b) The corresponding hinged dissection

J. Akiyama, M. Kano, and T. Sakai (Eds.): TJJCCGG 2012, LNCS 8296, pp. 1–19, 2013.

A pair of polygons α and β is said to be **equidecomposable** if α has a dissection into a finite number of pieces which can be rearranged to form β. Bolyai [2] and Gerwien [3] proved the following theorem:

Theorem A. ([2, 3]) *A pair of polygons P and Q is equidecomposable if and only if P and Q have the same area.*

Equidecomposability of a pair of polygons and polyhedra has been completely settled in [2] to [5].

Reversibility is a general case of Dudeney's hinged dissection and a specific case of equidecomposability. Related researches including equidecomposability, folding and unfolding problems, and Dudeney's puzzle can be found in [6] to [10].

Throughout this paper, unless explicitly stated, the polygons that we deal with will be convex polygons. In this paper, we dissect polygons along the edges of a tree-like structure that we refer to as a **dissection tree**. The hexagon in Fig. 2 is dissected along the edges of the dotted dissection tree.

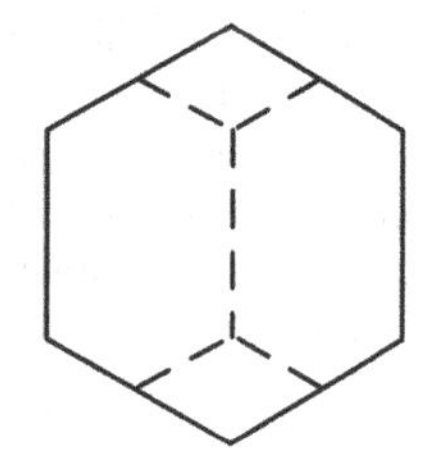

Fig. 2. An example of a dissection tree of a regular hexagon

A given pair of polygons α and β is said to be **reversible** if α and β have dissections into a common finite number of pieces along edges of dissection trees which can be rearranged to form β and α respectively, under the following conditions:

(i) The whole perimeter of one polygon fits into the interior of the other without gaps or overlaps and

(ii) The dissection tree of either polygon does not include any vertex of that polygon.

A polygon α is said to be **reversible** if there exists a polygon β such that the pair α and β is reversible.

When a pair α and β is reversible, we **hinge** the pieces of α (β) like a tree along the perimeter of α (β). It is always possible to hinge the n pieces of α (β) by using point hinges at arbitrary $n-1$ points among the n points on the perimeter of α (β) and transform the hinged pieces to both α and β continuously. Therefore we use models of hinged pieces without explicitly describing how to hinge and move the pieces.

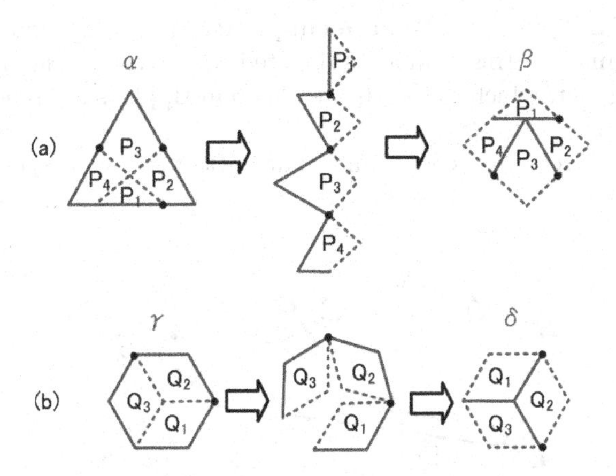

Fig. 3. (a) The solid triangle α and the dashed quadrilateral β give an example of reversible pair. (b) The pair of the solid hexagon γ and the dashed hexagon δ violates condition (ii) in the definition of reversibility.

Note that a pair of polygons with the same area is not always reversible but is always equidecomposable (Fig. 4).

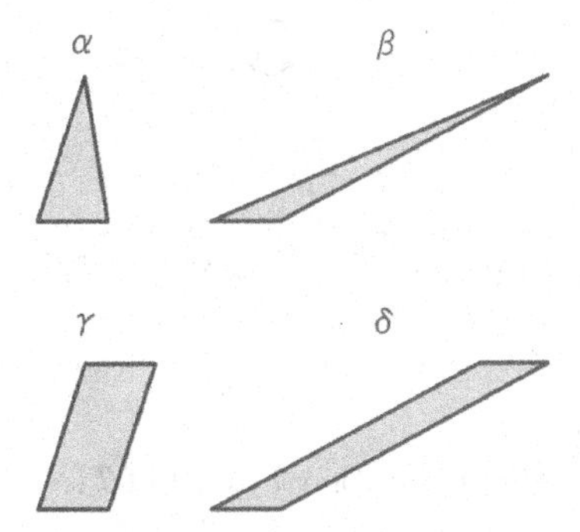

Fig. 4. Neither the pair α and β nor the pair γ and δ is reversible but each pair is equidecomposable

When the 2-dimensional plane is tiled by congruent copies of a polygon α without gaps nor overlaps, we say that α **tiles** the plane and denote the tiling by $\boldsymbol{T(\alpha)}$.

The following theorem is proved in [11].

Theorem B. ([11, Theorem 3.2]) *If a pair of polygons α and β is reversible, each of α and β tiles the plane by 180-degree rotations and translations only.*

The proof of Theorem B uses the fact that if a pair α and β is reversible, two superimposed tilings $T(\alpha)$ and $T(\beta)$ are obtained on the same plane by repeated reversals between α and β. The **superimposition** of $T(\alpha)$ and $T(\beta)$ on the same plane obtained in this manner is denoted by $\boldsymbol{T(\alpha, \beta)}$ (Fig. 5).

The following fact, which follows from Theorem B, plays an important role in further discussion:

The dissection trees between α and β are attained by the superimposition $T(\alpha, \beta)$.

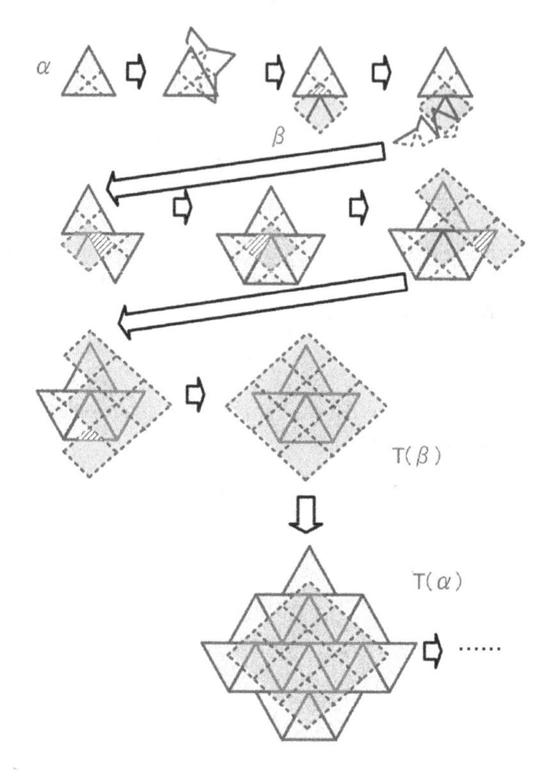

Fig. 5. The superimposition $T(\alpha, \beta)$ of $T(\alpha)$ and $T(\beta)$

2 Classification of Reversible Polygons

A convex polygon which tiles the plane by translations only is called a **parallelogon**. The set of all parallelogons is classified into two families, namely,

(i) parallelograms and

(ii) convex hexagons with three pairs of parallel sides, with each pair having the same length (Fig. 6).

The hexagons described in (ii) are called **parallelohexagons**. Denote the set of all parallelohexagons by $\boldsymbol{PH}$.

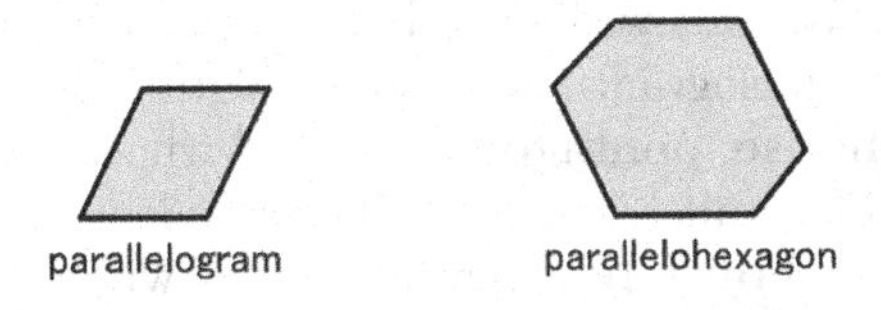

Fig. 6. Two families of parallelogons

A convex polygon which tiles the plane by 180-degree rotations and translations only is called a **quasi-parallelogon**. Denote the set of all quasi-parallelogons by $\boldsymbol{QP}$. Parallelogons are included in the set $\boldsymbol{QP}$. The set $\boldsymbol{QP}$ can be classified with respect to the number of sides of a polygon into four **families**, namely,

(i) triangles;

(ii) convex quadrilaterals;

(iii) **quasi-parallel pentagons**, which are convex pentagons with at least one pair of parallel sides; and

(iv) **quasi-parallel hexagons**, which are convex hexagons with at least one pair of parallel opposite sides with the same length (Fig. 7).

Denote the set of all quasi-parallel pentagons and the set of all quasi-parallel hexagons by $\boldsymbol{QPP}$ and $\boldsymbol{QPH}$, respectively.

A pair of parallel opposite sides of a polygon $P \in \boldsymbol{QPH}$ with the same length is called a **zone** of P. Note that P has either 1 or 3 zones and that P belongs to $\boldsymbol{PH}$ if and only if $P \in \boldsymbol{QPH}$ and P has 3 zones. Thus $\boldsymbol{PH}$ is a subset of $\boldsymbol{QPH}$.

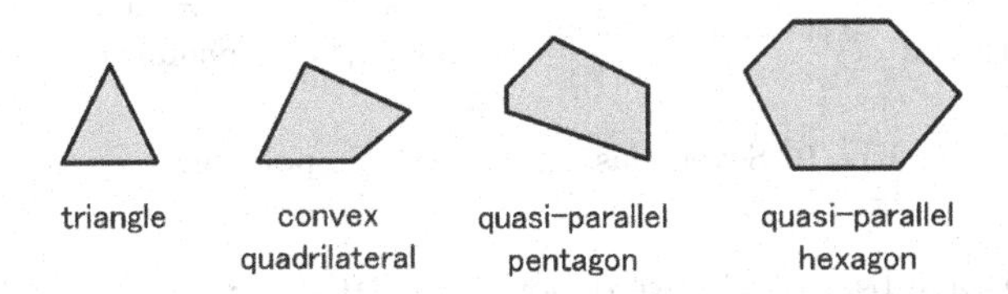

Fig. 7. Four families of quasi-parallelogons

For later discussion, it is convenient to divide the family $\boldsymbol{QPP}$ into two subfamilies depending on whether there exists a pair of parallel sides with the same length or not. We also divide the family of convex quadrilaterals into three families depending on the number of **zones**, which are pairs of parallel sides (Fig. 8, see [14] for details).

Note that the term "zone" is used in two different ways depending on whether the polygon is a hexagon or not.

Recall that every reversible polygon belongs to the set $\boldsymbol{QP}$ by Theorem B. Each reversible polygon belongs to one of the seven **classes**, namely,

(i) the set of all triangles, denoted by $\mathfrak{P}_1$;

(ii) the set of all convex quadrilaterals with no zones, denoted by $\mathfrak{P}_2$;

(iii) the set of all trapezoids with exactly one zone, denoted by $\mathfrak{P}_3$;

(iv) the set of all parallelograms, denoted by $\mathfrak{P}_4$;

(v) the set of all **house pentagons**, each of whose elements is a convex pentagon with a pair of parallel sides with the same length, denoted by $\mathfrak{P}_5$;

(vi) the set of all **non-house pentagons**, each of whose elements is a convex pentagon with at least one pair of parallel sides with different lengths, denoted by $\mathfrak{P}_6$; and

(vii) the set of all quasi-parallel hexagons, denoted by $\mathfrak{P}_7$ (Fig. 9).

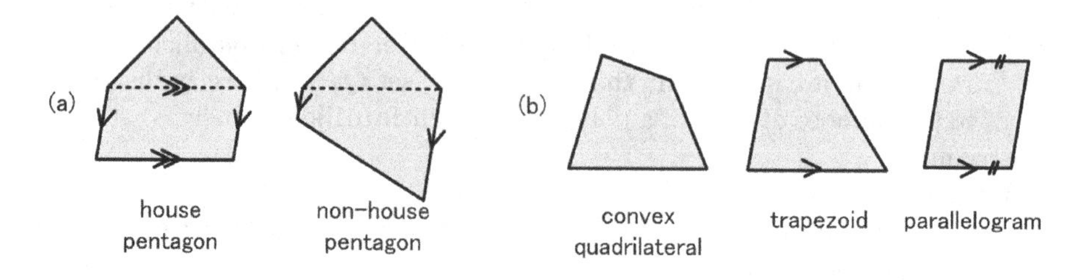

Fig. 8. (a) House and non-house pentagon (b) Quadrilateral with 0, 1, and 2 zones

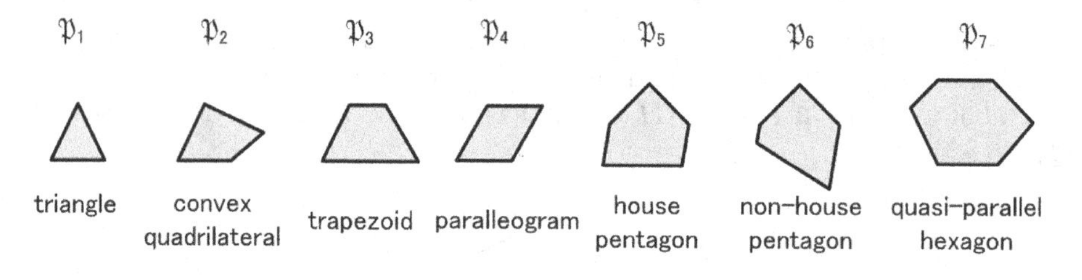

Fig. 9. Seven classes of reversible polygons

Since $\boldsymbol{PH}$ is a subset of $\mathfrak{P}_7$, we denote it by $\mathfrak{P}'_7$.

Consider a polygon $P \in \boldsymbol{QP}$. Let $\boldsymbol{P'}$ be its **half turn** (180° rotation of P). A figure obtained by gluing P and P' in such a way that they share (a part of) an edge is called a **concatenation** of P. Denote it by $\boldsymbol{c(P)}$. Note that there are infinitely many possible concatenations of P (Fig. 10).

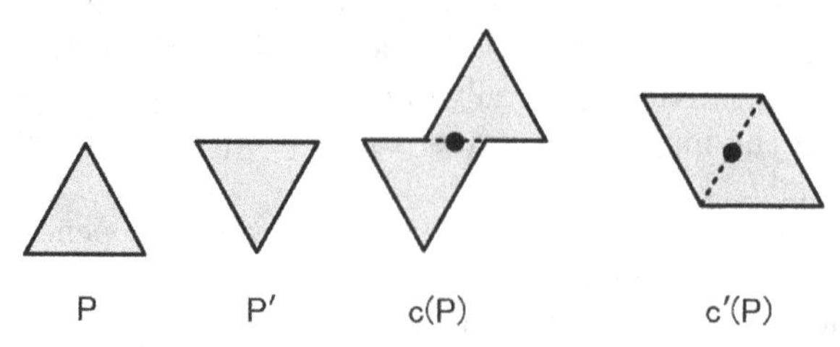

Fig. 10. A triangle P, its half turn P', a concatenation $c(P)$, and another concatenation $c'(P)$

3 Operators for Reversible Polygons

3.1 Stretching Operators

Consider a polygon $P = ABCDEF \in \mathfrak{P}_7$ with a zone with $AB \parallel ED$ and $\overline{AB} = \overline{ED}$. Denote the centers of BC, CD, EF, and FA by G, H, I, and J, respectively. The parallelogram $GHIJ$ is called a **reversion trunk $R(P)$** of P (Fig. 11).

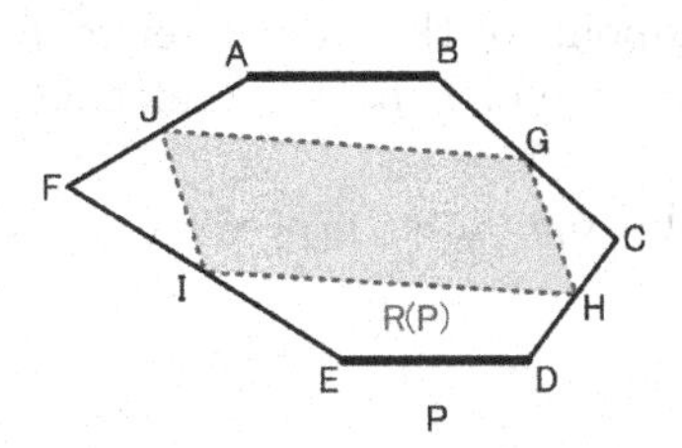

Fig. 11. Reversion trunk $R(P)$ (gray parallelogram)

Conversely, let R be a parallelogram and let X and Y be two distinct points in R such that the line passing through X and Y intersects a pair of parallel sides of R. The pair of parallel sides of R, which has an intersection with the line XY, is called a **non-zonal pair**. The other pair of parallel sides of R is called a **zonal pair**.

A quasi-parallel hexagon is constructed from R as follows:

(i) Reflect X through the midpoint of the side closer to X between the non-zonal pair. Denote the image of X by C.

(ii) Reflect Y through the midpoint of the side closer to Y between the non-zonal pair. Denote the image of Y by F.

(iii) Reflect C through each of the two terminal points of the side closer to X between the non-zonal pair. Denote the images of C by B and D, respectively. Reflect F through each of the two terminal points of the side closer to Y between the non-zonal pair. Denote the images of F by E and A, respectively.

Note that the hexagon $ABCDEF$, which depends on X and Y, belongs to $\mathfrak{P}_7$ and has R as one of its reversion trunks.

Given $ABCDEF(X,Y), A'B'C'D'E'F'(X',Y') \in \mathfrak{P}_7$, an operator $f_{R,X,Y,X',Y'}$ is called a **stretching operator** if $f_{R,X,Y,X',Y'}(ABCDEF(X,Y)) = A'B'C'D'E'F'(X',Y')$ (Fig. 12).

In the discussion below, we simply denote by f a stretching operator $f_{R,X,Y,X',Y'}$.

Lemma 1. *For arbitrary polygons $P, Q \in \mathfrak{P}_7$, the following two propositions are equivalent:*

(i) There exists a stretching operator f such that $P = f(Q)$.

(ii) The polygons P and Q have a common reversion trunk.

Proof. (1) (i) $\Rightarrow$ (ii)

Suppose that there exists a stretching operator f such that $P = f(Q)$.

There exists a parallelogram R which is a common reversion trunk $R = R(P) = R(Q)$ by the definition of stretching operator.

(2) (ii) $\Rightarrow$ (i)

Suppose that there exist reversion trunks $R(P)$ and $R(Q)$ such that $R(P) = R(Q)$.

Let C and F be two vertices of P which are not included in the zone of P corresponding to $R(P)$, respectively. Reflect C through the midpoint of the side closer to C between the non-zonal pair of $R(P)$. Denote the image of C by X. Reflect F through the midpoint of the side closer to F between the non-zonal pair of $R(P)$. Denote the image of F by Y. Analogously, we obtain X' and Y' from Q.

We let f be the stretching operator $f_{R(P),X,Y,X',Y'}$ (Fig. 12).

□

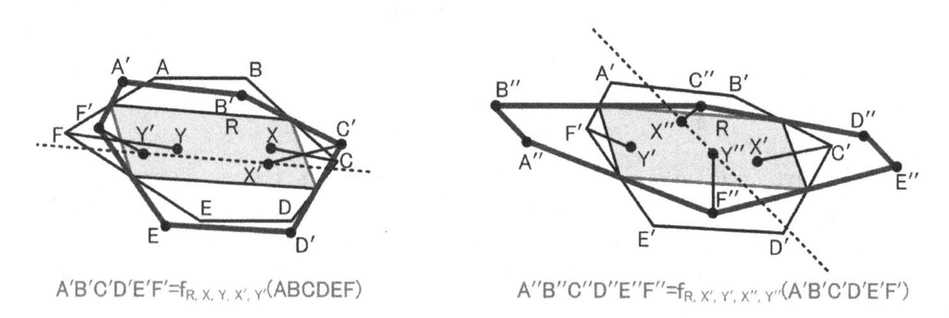

Fig. 12. Two examples of stretching operators with different zonal pairs

We extend stretching operators to the set $\boldsymbol{S}$ of all convex quadrilaterals other than parallelograms and all pentagons, i.e., $\boldsymbol{S} \equiv \mathfrak{P}_2 \cup \mathfrak{P}_3 \cup \mathfrak{P}_5 \cup \mathfrak{P}_6$. Take an element $P \in \boldsymbol{S}$ and consider some concatenation (not necessarily unique) $c(P) \in \mathfrak{P}_7$. We denote by $f_{c(P)}$ the extension of a stretching operator f on $P \in \boldsymbol{S}$ which is defined to satisfy that $c(f_{c(P)}(P)) = f(c(P))$

When $P \in \boldsymbol{S} \setminus \mathfrak{P}_6$, $f_{c(P)}(P)$ is determined by the equation $c(f_{c(P)}(P)) = f(c(P))$. When $P = AXYEF \in \mathfrak{P}_6$, we have $c(P) = ABCDEF \in \mathfrak{P}_7$ with a zone with $\overline{AB} = \overline{ED}$ and $AB \parallel ED$. Then $f_{c(P)}(P) = A'X'Y'E'F'$ is determined by cutting $f(c(P)) = A'B'C'D'E'F'$ so that the internal ratio of the cut zone of $c(P)$ is preserved, i.e., $\overline{AX} : \overline{XB} : \overline{DY} : \overline{YE} = \overline{A'X'} : \overline{X'B'} : \overline{D'Y'} : \overline{Y'E'}$ (Fig. 13). In the discussion below, we simply denote by f a stretching operator $f_{c(P)}$.

3.2 Cutting Operators

We consider some concatenation $c(P) \in \mathfrak{P}'_7$ (not necessarily unique) of a polygon $P \in \mathfrak{P}_6$ and call the glued edge e and its midpoint O. A non-house pentagon, say $Q \in \mathfrak{P}_6$, is obtained by cutting $c(P)$ along some line e' passing through O and intersecting the same zone of $c(P)$ as the zone intersecting e. An operator $f_{P,Q}$ is called a **cutting operator** if $f_{P,Q}(P) = Q$ (Fig. 14).

In the succeeding discussion, we simply denote a cutting operator $f_{P,Q}$ by f.

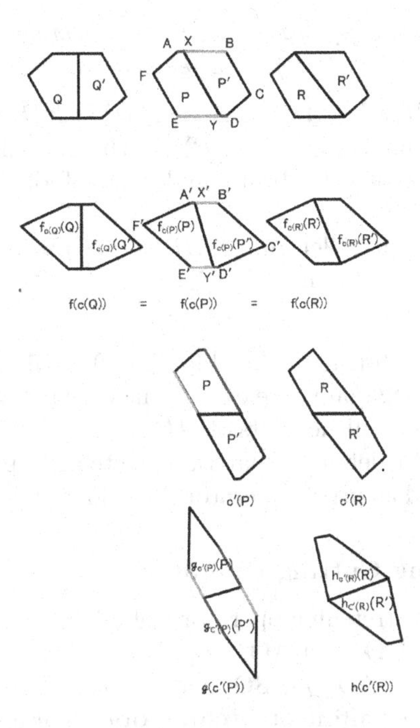

Fig. 13. Extensions $f_{c(P)}$, $f_{c'(P)}$, $f_{c(Q)}$, $g_{c'(P)}$, and $h_{c'(R)}$ of stretching operators f, g, and h, where $P \in \mathfrak{P}_6$, $Q \in \mathfrak{P}_5$, and $R \in \mathfrak{P}_2 \cup \mathfrak{P}_3$

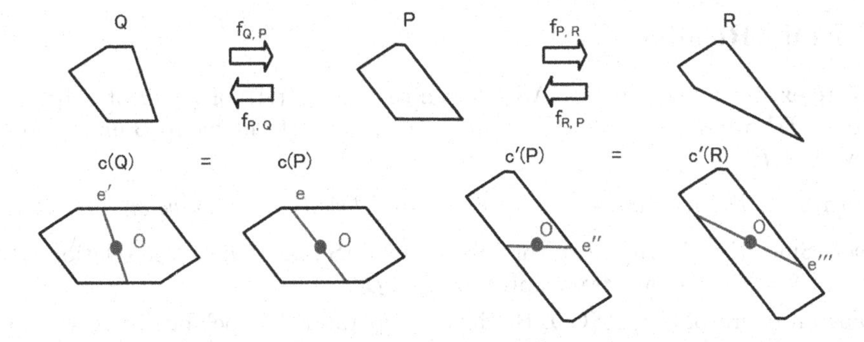

Fig. 14. Two non-house pentagons Q and R attained by two cutting operators based on the concatenations $c(P)$ and $c'(P)$, corresponding to each pair of parallel sides, respectively

3.3 Affine Operators

An **affine operator** is a composition of a linear operator and a translation operator. Note that affine operators on a plane preserve convexity, parallelism, and internal ratio. We can perform an affine operator f on polygons, tilings, concatenations, trunks, etc.

Lemma 2. *For an arbitrary polygon $P \in \mathfrak{P}_7$, a reversion trunk $R(P)$, and an arbitrary affine operator f, $f(R(P))$ is a reversion trunk $R(f(P))$.*

Proof. Let $P = ABCEF$ with $AB \parallel ED$ and $\overline{AB} = \overline{ED}$. Since f preserves the internal ratio and the four vertices of $R(P)$ are the midpoints of BC, CD, EF, and FA, the four vertices of $f(R(P))$ are midpoints of four sides $f(BC)$, $f(CD)$, $f(EF)$, and $f(FA)$ of $f(P)$.

Since f preserves the parallelism, $f(AB) \parallel f(ED)$ and $\overline{f(AB)} = \overline{f(ED)}$. Thus $f(R(P))$ is a reversion trunk $R(f(P))$.

□

So far, stretching operators are defined for $\mathfrak{P}_2 \cup \mathfrak{P}_3 \cup \mathfrak{P}_5 \cup \mathfrak{P}_6 \cup \mathfrak{P}_7$ and cutting operators are defined for $\mathfrak{P}_6$ only. We extend the domain of stretching operators and of cutting operators to all elements of $\boldsymbol{QP}$ by assigning the identity operator when any operator is not defined. That is, a stretching operator is the identity operator for $\mathfrak{P}_1 \cup \mathfrak{P}_4$ and a cutting operator is the identity operator for $\boldsymbol{QP} \setminus \mathfrak{P}_6$.

3.4 Affine Stretching Cutting Operators

We denote the set of all stretching operators, of all cutting operators, and of all affine operators by **SO**, **CO**, and **AO**, respectively.

Consider operators $f \in$ AO, $g \in$ SO, and $h \in$ CO. The composite operators $g \circ f$ and $h \circ g \circ f$ are called **affine stretching operator** and **affine stretching cutting operator**, respectively. Denote the set of all affine stretching operators and of all affine stretching cutting operators by **ASO** and **ASCO**, respectively.

4 Main Results

If there exists an operator $f \in$ ASCO such that $B = f(A)$ for a pair of polygons $A, B \in \boldsymbol{QP}$, we say A is affine stretching cutting transformable to B and denote it by $\boldsymbol{A \equiv B}$.

Lemma 3. *The relation $\equiv$ among quasi-parallelogons is an equivalence relation.*

Proof. Since the identity operator i belongs to ASCO, we have the relation $A \equiv i(A) = A$ for an arbitrary polygon $A \in \boldsymbol{QP}$.

For an operator $f \in$ ASCO, the inverse operator f^{-1} belongs to ASCO by the definitions of AO, SO, and CO.

Suppose the relation $A \equiv B$ holds for polygons $A, B \in \boldsymbol{QP}$, then there exists an operator $f \in$ ASCO such that $A = f(B)$. Since $B = f^{-1}(A)$, we have the relation $B \equiv A$.

For operators $f, g \in$ ASCO, the composite operator $g \circ f$ belongs to ASCO by the definitions of AO, SO, and CO.

Suppose the relations $A \equiv B$ and $B \equiv C$ holds for polygons $A, B, C \in \boldsymbol{QP}$. Since there exist operators $f, g \in$ ASCO such that $A = f(B) = f(g(C)) = g(f(C))$, we have the relation $A \equiv C$.

Thus, the relation defined is an equivalence relation.

□

Theorem 1. *The seven classes $\mathfrak{P}_1, \mathfrak{P}_2, \cdots, \mathfrak{P}_7$ of quasi-parallelogons are equivalence classes under the relation $\equiv$, i.e., polygons A and B satisfy the relation $A \equiv B$ if and only if there exists a class $\mathfrak{P}_i$ $(i = 1, 2, \cdots, 7)$ such that $A, B \in \mathfrak{P}_i$.*

Proof. First, suppose that there exists a class $\mathfrak{P}_i$ $(i = 1, 2, \cdots, 7)$ such that $A, B \in \mathfrak{P}_i$.

Case 1: $i = 1$

There exist concatenations $c(A), c(B) \in \mathfrak{P}_4$ and an operator $f \in$ AO such that $c(A) = f(c(B)) = c(f(B))$. Then $A = f(B)$ holds.

Case 2: $i = 2, 3, 5$

There exist concatenations $c(A), c(B) \in \mathfrak{P}'_7$ and an operator $f \in$ AO such that $R(c(A)) = f(R(c(B)))$, where $R(c(A))$ and $R(c(B))$ are reversion trunks of $c(A)$ and $c(B)$, respectively. By Lemma 2, $R(c(A)) = f(R(c(B))) = R(f(c(B))) = R(c(f(B)))$. By Lemma 1, there exists an operator $g \in$ SO such that $c(A) = g(c(f(B))) = c(g(f(B)))$. Then $A = f(g(B)) = g(f(B))$ holds.

Case 3: $i = 4$

There exists an operator $f \in$ AO such that $A = f(B)$.

Case 4: $i = 6$

There exist concatenations $c(A), c(B) \in \mathfrak{P}'_7$ and an operator $f \in$ AO where $R(c(A)) = f(R(c(B)))$, since $R(c(A))$ and $R(c(B))$ are reversion trunks of $c(A)$ and $c(B)$, respectively.

By Lemma 2, $R(c(A)) = f(R(c(B)) = R(f(c(B)) = R(c(f(B)))$.

By Lemma 1, there exists an operator $g \in$ SO such that $c(A) = g(c(f(B)) = c(g(f(B)))$. Thus there exits an operator $h \in$ CO such that $A = h(g(f(B)))$ holds.

Case 5: $i = 7$

There exists an operator $f \in$ AO such that $R(A) = f(R(B))$. By Lemma 2, we have $R(A) = f(R(B)) = R(f(B))$. By Lemma 1, there exists an operator $g \in$ SO such that $A = g(f(B))$ holds.

In each case, the relation $A \equiv B$ holds.

Conversely, suppose that $A \equiv B$. By the definition of the relation $\equiv$, there exists an operator $f \in$ ASCO such that $B = f(A)$. Let $\mathfrak{P}_i$ be the class to which B belongs. Since f preserves the number of edges of a polygon, A has the same number of edges as B.

Since a cutting operator is the identity operator for convex quadrilaterals and an affine stretching operator preserves the number of zones, we have $A \in \mathfrak{P}_2$ if $B \in \mathfrak{P}_2$, $A \in \mathfrak{P}_3$ if $B \in \mathfrak{P}_3$, and $A \in \mathfrak{P}_4$ if $B \in \mathfrak{P}_4$.

Since a cutting operator is the identity operator for house pentagons and an affine stretching operator preserves the ratio of the lengths of a pair of parallel sides, we have $A \in \mathfrak{P}_5$ if $B \in \mathfrak{P}_5$ and $A \in \mathfrak{P}_6$ if $B \in \mathfrak{P}_6$.

Therefore A belongs to $\mathfrak{P}_i$.

□

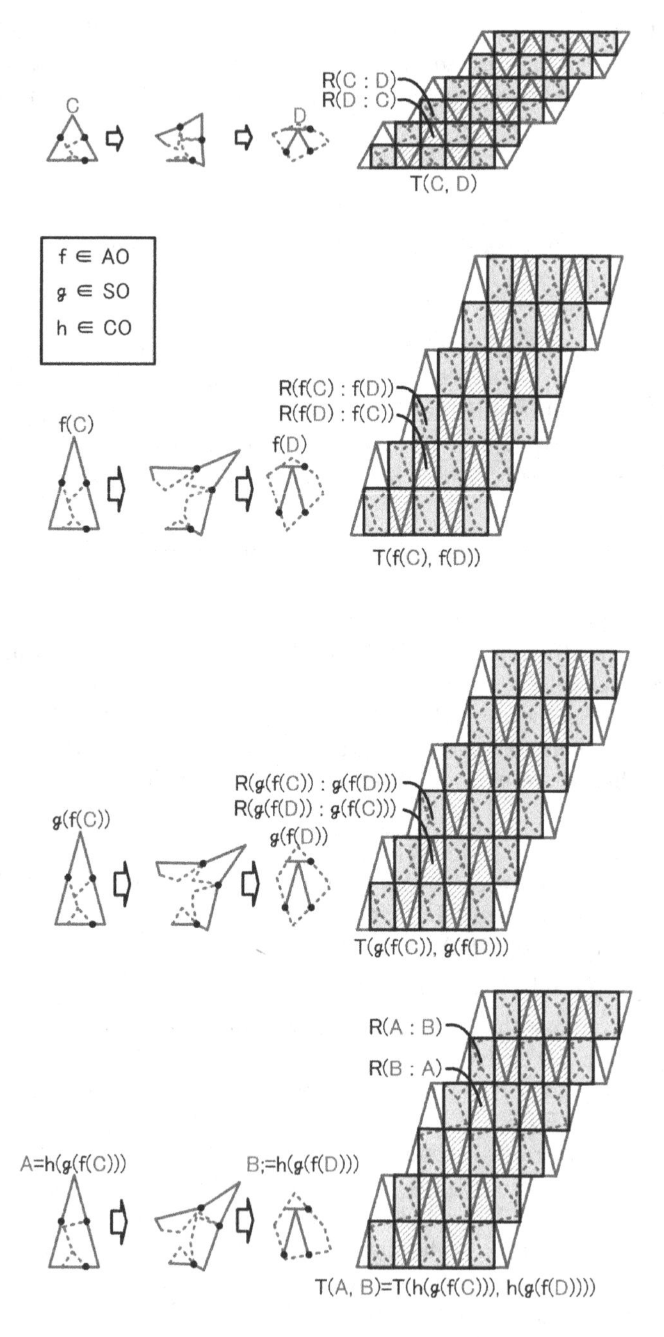

Fig. 15. A procedure to attain the dissection trees between $A = h(g(f(C)))$ and $B = h(g(f(D)))$, the superimposition $T(A, B)$, and the checkerboard consisting of $R(A : B)$'s and $R(B : A)$'s

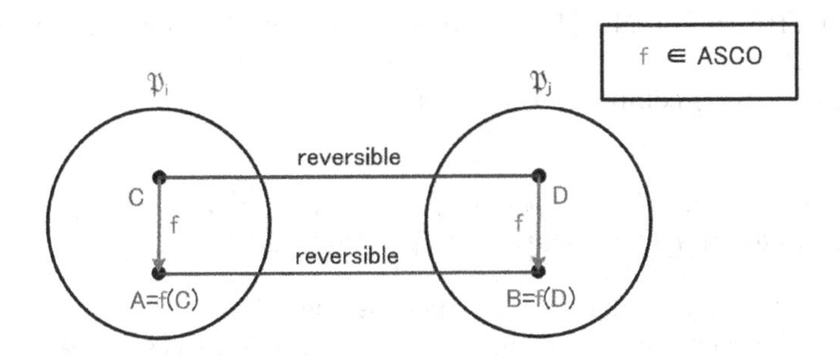

Fig. 16. An affine stretching cutting operator preserves reversibility in the sense of Theorem 2

In Section 3.1, we defined reversion trunks $R(P)$ for a polygon $P \in \mathfrak{P}_7$.

We now extend this notion to every $P \in \boldsymbol{QP}$. Given a reversible polygon P, consider some polygon Q such that the pair P and Q is reversible. We define a **reversion trunk $\boldsymbol{R(P : Q)}$** of P as the convex hull of the terminal points of a dissection tree $T(Q) \cap P$ of P. The reversion trunk $R(P : Q)$ depends on the choice of Q. We denote by $\boldsymbol{\mathfrak{R}(P)}$ the set of all reversion trunks $R(P : Q)$ where the pair P and Q is reversible.

Recall that for a reversible pair P and Q, the dissection tree between P and Q is induced by the superimposition $T(P, Q)$.

Note that in the superimposition $T(P, Q)$, reversion trunks $R(P : Q)$ and $R(Q : P)$ appear alternatively, like black parts and white parts in a checkerboard, respectively. Thus we have the equation $R(P : Q) = R(Q : P)$ and all the reversion trunks of P and Q are parallelograms. This fact guarantees that our definitions of reversion trunks for a member of $\boldsymbol{QP}$ is a generalization of the definition for a member of $\mathfrak{P}_7$. See [13] for details.

In order to study reversibility among quasi-parallelogons, we prove the following theorem. An analogous result for reversibility among 3-dimensional parallelohedra is obtained in [12].

Theorem 2. *Suppose that a pair $C \in \mathfrak{P}_i$ and $D \in \mathfrak{P}_j$ $(1 \leq i, j \leq 7)$ is reversible. Then, for an arbitrary polygon $A \in \mathfrak{P}_i$, there exists a polygon $B \in \mathfrak{P}_j$ such that the pair A and B is reversible.*

Proof. The dissection tree between C and D is induced by the superimposition $T(C, D)$. Let $R(C : D)$ and $R(D : C)$ be the reversion trunks attained by $T(C, D)$.

Consider an arbitrary polygon $A \in \mathfrak{P}_i$.

Since A and C belong to the same class $\mathfrak{P}_i$, there exists an operator $f \in$ ASCO such that $A = f(C)$ by Theorem 1.

Note that every affine stretching operator preserves the number of edges, the convexity of each of the pieces, and the number of zones of polygons. Thus the dissection tree between $f(C)$ and $f(D)$ is induced by $T(f(C), f(D))$ (Fig. 15).

The pair $f(C) = A$ and $f(D)$ is reversible, where $f(D)$ belongs to the same class $\mathfrak{P}_j$ as D.

Letting $B \equiv f(D)$ completes the proof.

□

Theorem 3. *For an arbitrary pair of families* $\mathfrak{P}_i$ *and* $\mathfrak{P}_j$ *(*$1 \leq i, j \leq 7$*), there exists a reversible pair* $C \in \mathfrak{P}_i$ *and* $D \in \mathfrak{P}_j$.

Proof. Without loss of generality, we may assume that $i \leq j$.

We prove the theorem by showing concrete dissection trees for each case.

For $(i, j) = (1, 1)$

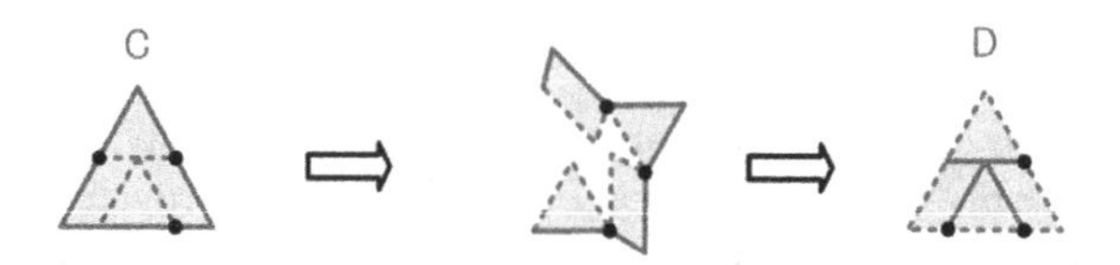

For $(i, j) = (1, 2)$

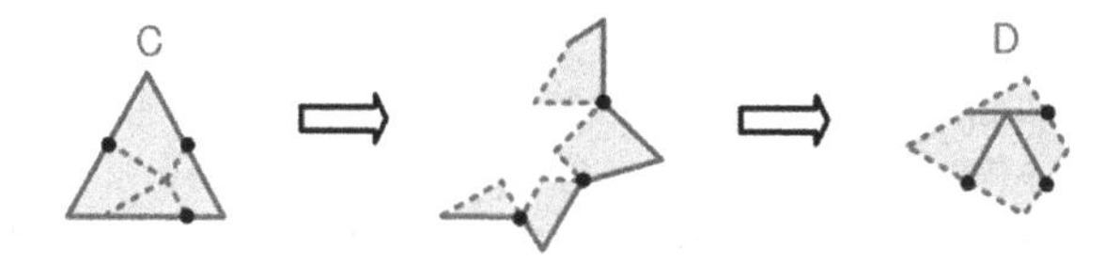

For $(i, j) = (1, 3)$

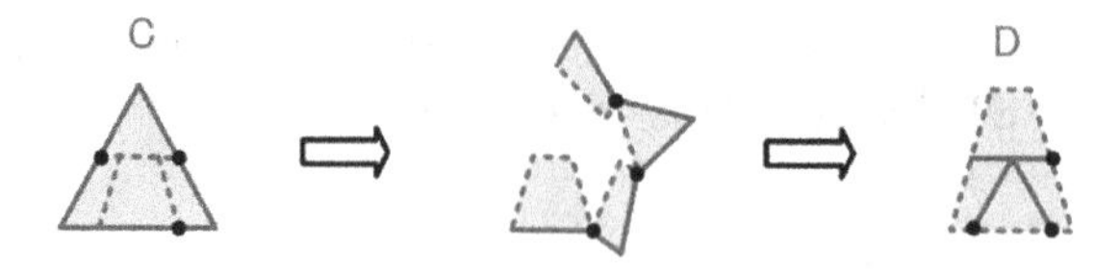

For $(i, j) = (1, 4)$

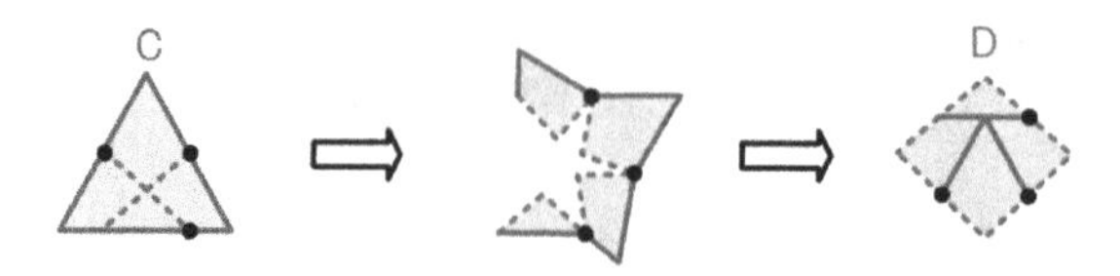

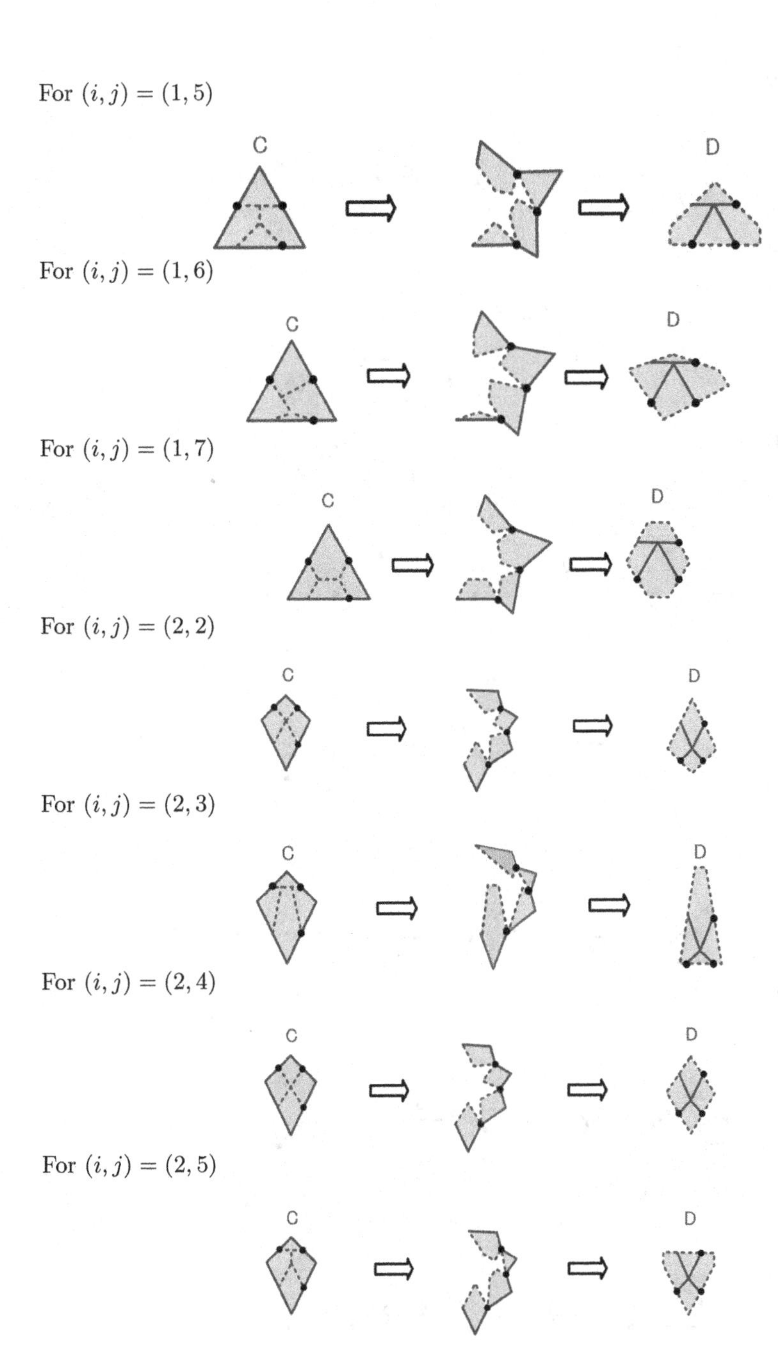
For (i, j) = (1, 5)
C
D
For (i, j) = (1, 6)
C
D
For (i, j) = (1, 7)
C
D
For (i, j) = (2, 2)
C
D
For (i, j) = (2, 3)
C
D
For (i, j) = (2, 4)
C
D
For (i, j) = (2, 5)
C
D

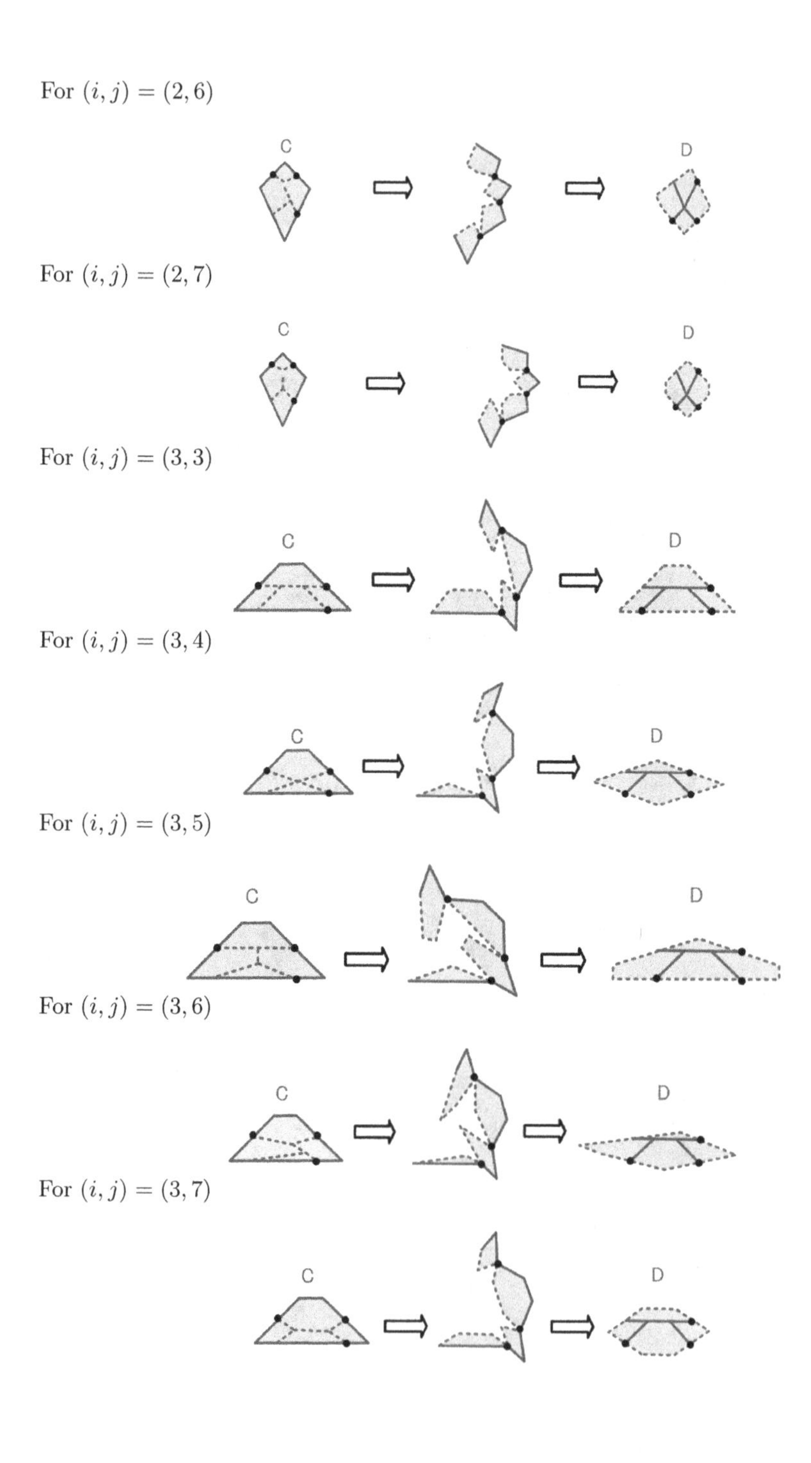
For $(i,j) = (2,6)$
C
D
For $(i,j) = (2,7)$
C
D
For $(i,j) = (3,3)$
C
D
For $(i,j) = (3,4)$
C
D
For $(i,j) = (3,5)$
C
D
For $(i,j) = (3,6)$
C
D
For $(i,j) = (3,7)$
C
D

For $(i, j) = (4, 4)$

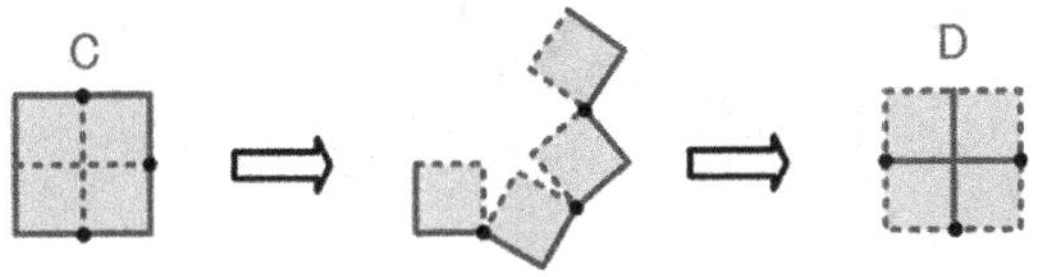

For $(i, j) = (4, 5)$

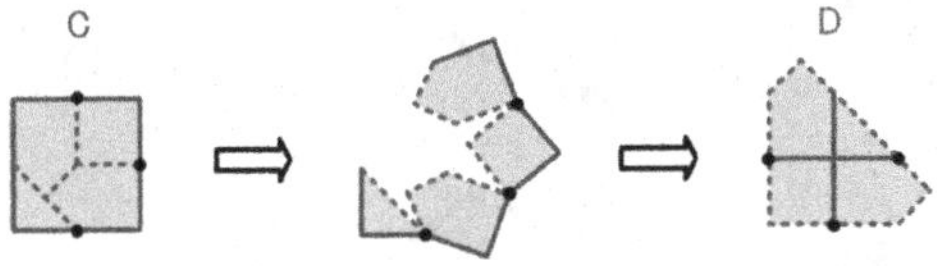

For $(i, j) = (4, 6)$

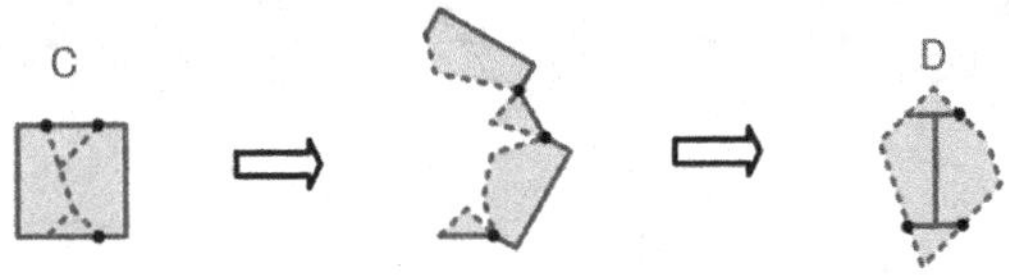

For $(i, j) = (4, 7)$

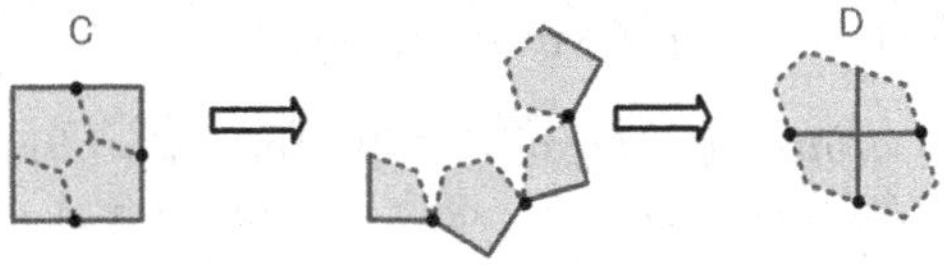

For $(i, j) = (5, 5)$

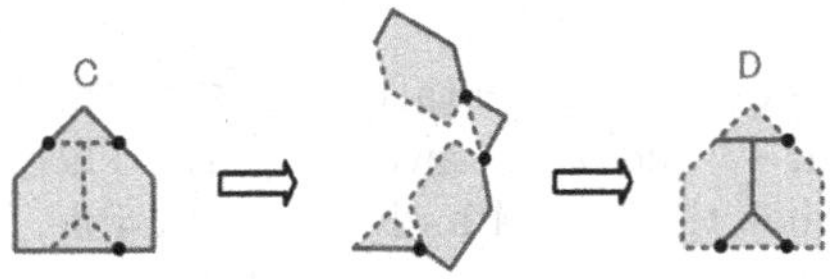

For $(i, j) = (5, 6)$

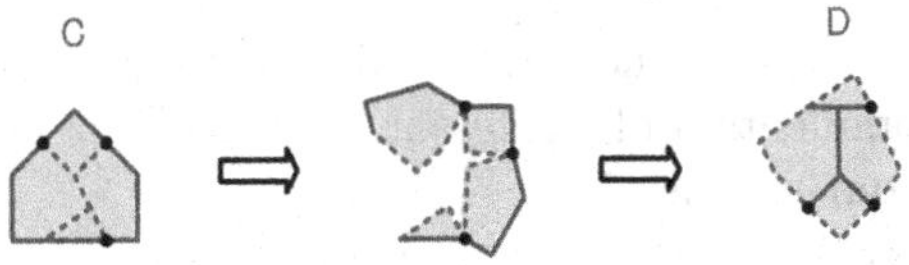

For $(i, j) = (5, 7)$

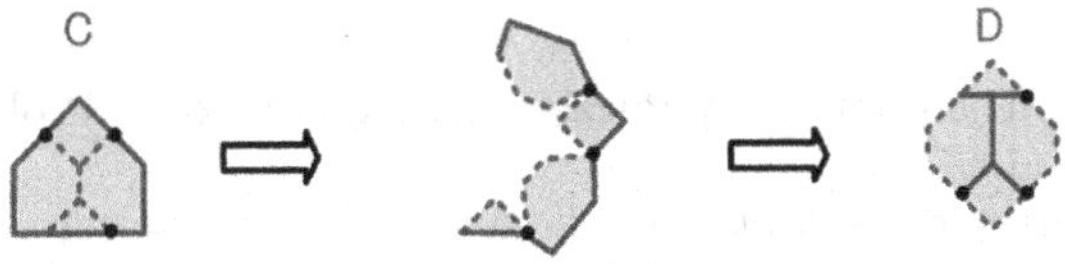

For $(i, j) = (6, 6)$

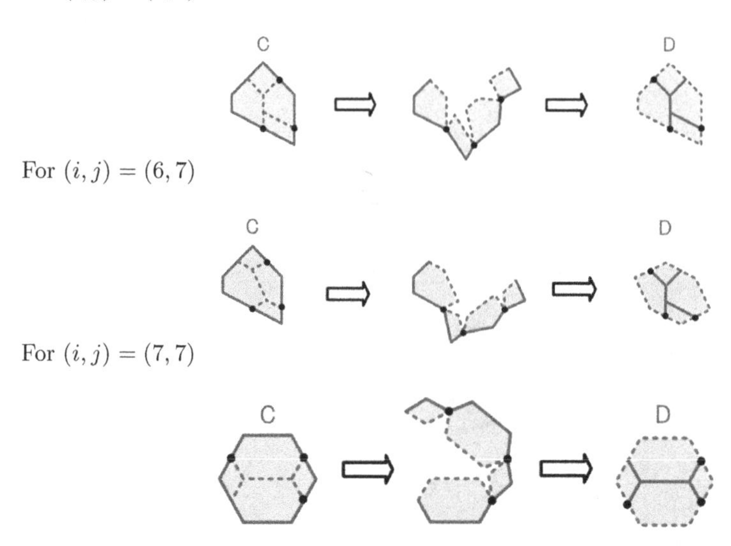

□

Let us call the polygons C and D that appear in the proof of Theorem 3 the **representatives** of the cases (i, j), respectively.

Theorem 4. *For an arbitrary polygon* $A \in \mathfrak{P}_i$ *and an arbitrary family* $\mathfrak{P}_j$ $(1 \leq i, j \leq 7)$*, there exists a polygon* $B \in \mathfrak{P}_j$ *such that the pair* A *and* B *is reversible.*

Proof. Consider an arbitrary polygon $A \in \mathfrak{P}_i$ and an arbitrary family $\mathfrak{P}_j$ $(1 \leq i, j \leq 7)$.

By Theorem 3, there exist representatives $C \in \mathfrak{P}_i$ and $D \in \mathfrak{P}_j$ of the case (i, j) such that the pair C and D is reversible. By Theorem 2, there exists a polygon $B \in \mathfrak{P}_j$ such that the pair A and B is reversible.

□

Acknowledgements. The authors would like to thank a referee for giving us various important comments and Professor Mari-Jo Ruiz and Professor Agnes Garciano for editing our manuscript.

References

1. Dudeney, H.E.: The Canterbury Puzzles and Other Curious Problems. W. Heinemann, London (1907)
2. Bolyai, F.: Tentamen juventutem studiosam in elementa matheseos puræ, elementaris ac sublimioris, methodo intuitiva, evidentiaque huic propria, introducendi, cum appendice triplici. Typis Collegii Reformatorum per Josephum, et Simeonem kali de felső Vist. Maros Vásárhelyini (1832-1833)

3. Gerwien, P.: Zerschneidung jeder beliebigen Anzahl von gleichen geradlinigen Figuren in dieselben Stücke. In: Crelle, A.L. (ed.) Journal für Die Reine und Angewandte Mathematik, vol. 10, pp. 228–234. Taf. III. Walter de Gruyter, Berlin (1833)
4. Hilbert, D.: Mathematische Probleme. Vortrag, gehalten auf dem internationalen Mathematiker-Kongreß zu Paris 1900. In: Horstmann, L. (ed.) Nachrichten von der Königl. Gesellschaft der Wissenschaften zu Göttingen, Mathematisch-Physikalische Klasse, vol. 1900, pp. 253–297. Commissionverlag der Dieterich'schen Universitätsbuchhandlung, Göttingen (1900); Subsequently in Bulletin of the American Mathematical Society, vol. 8(10), pp. 437–479. American Mathematical Society, New York (1902)
5. Dehn, M.: Über den Rauminhalt. In: Klein, F., Dyck, W., Hilbert, D. (eds.) Mathematische Annalen, vol. 55(3), pp. 465–478. H. G. Teubner, Leipzig (1901)
6. Frederickson, G.N.: Dissections: Plane and Fancy. Cambridge University Press, New York (1997)
7. Frederickson, G.N.: Hinged Dissections: Swinging and Twisting. Cambridge University Press, New York (2002)
8. Frederickson, G.N.: Piano-Hinged Dissections: Time to Fold. A K Peters, Wellesley (2006)
9. Boltyanskii, V.G.: Equivalent and Equidecomposable Figures. D. C. Health and Co., Lexington (1963); translated and adapted from the first Russian edition by Henn, A.K., Watts, C.E (1956)
10. Dolbilin, N., Itoh, J., Nara, C.: Affine Classes of 3-Dimensional Parallelohedra - Their Parametrizations (in this volume)
11. Akiyama, J., Nakamura, G.: Congruent Dudeney Dissections of Triangles and Convex Quadrilaterals — All Hinge Points Interior to the Sides of the Polygons. In: Aronov, B., Basu, S., Pach, J., Sharir, M. (eds.) Discrete and Computational Geometry. Algorithms and Combinatorics, vol. 25, pp. 43–63. Springer, Heidelberg (2003)
12. Akiyama, J., Sato, I., Seong, H.: On Reversibility among Parallelohedra. In: Márquez, A., Ramos, P., Urrutia, J. (eds.) EGC 2011. LNCS, vol. 7579, pp. 14–28. Springer, Heidelberg (2012)
13. Akiyama, J., Seong, H.: A Criterion for Two Polygons to be Reversible (to be published)
14. Akiyama, J., Rappaport, D., Seong, H.: Decision Algorithm and Classification of Reversible Pairs of Polygons (to be published)

Colored Quadrangulations with Steiner Points

Victor Alvarez[1,⋆] and Atsuhiro Nakamoto[2,⋆⋆]

[1] Fachrichtung Informatik
Universität des Saarlandes
Saarbrücken, Germany
alvarez@cs.uni-saarland.de
[2] Department of Mathematics
Yokohama National University
Yokohama, Japan
nakamoto@ynu.ac.jp

Abstract. Let $P \subset \mathbb{R}^2$ be a k-colored set of n points in general position, where $k \geq 2$. A k-colored quadrangulation on P is a properly colored straight-edge plane graph G with vertex set P such that the boundary of the unbounded face of G coincides with $\mathrm{CH}(P)$ and that each bounded face of G is quadrilateral, where $\mathrm{CH}(P)$ stands for the boundary of the convex hull of P. It is easily checked that in general not every k-colored P admits a k-colored quadrangulation, and hence we need the use of *Steiner points*, that is, auxiliary points which can be put in any position of the interior of the convex hull of P and can have any color among the k colors. In this paper, we show that if P satisfies some condition for colors of the points in the convex hull, then a k-colored quadrangulation of P can always be constructed using less than $\frac{(16k-2)n+7k-2}{39k-6}$ Steiner points. Our upper bound improves the known upper bound for $k = 3$, and gives the first bounds for $k \geq 4$.

1 Introduction

Let $P \subset \mathbb{R}^2$ be a set of n points on the plane, a *point set* or an *n-point set* for short. Let $\mathrm{CH}(P)$ denote the boundary of the convex hull of P. We always assume that P is in general position, that is, no three points of P are collinear. A quadrangulation of P is a straight-edge plane graph G with vertex set P such that the boundary of the unbounded face of G coincides with the convex hull of P, and that every bounded face of G is quadrilateral.

Quadrangulations of a point set received an extensive attention back in the 90's, where they were sometimes preferred over triangulations in the study of finite element methods and scattered data interpolation, see [8] for example. It is not hard to see that *not every* P admits a quadrangulation. It can be verified that a necessary condition for a point set P to admit a quadrangulation is $|P| \geq 4$, and $\mathrm{CH}(P)$ must be even-sided. It turns out that these two conditions are also sufficient, see [3,9].

⋆ Partially supported by CONACYT-DAAD of México.
⋆⋆ Partially supported by JSPS KAKENHI Grant Number 21340119.

J. Akiyama, M. Kano, and T. Sakai (Eds.): TJJCCGG 2012, LNCS 8296, pp. 20–29, 2013.

How about quadrangulations of point sets with special properties? For example, each face of the quadrangulation is required to be a convex quadrilateral, see [2,4,6,10,11]. Even in this setting it was shown in [2] that again, not every set of points admits a *convex quadrangulation.* It was shown that one can always construct a convex quadrangulation using at most $\frac{3n}{2}$ *Steiner points*, that is, auxiliary points which can be put in any position of the interior of the convex hull of the given point set. On the other hand, $\frac{n}{4}$ are sometimes necessary [2]. Later both bounds were improved to roughly $\frac{5n}{4}$ and $\frac{n}{3}$ respectively [6].

Another kind of quadrangulations arises when a point set P is colored with $k \geq 2$ colors, and we look for a quadrangulation not containing *monochromatic* edges, that is, edges whose ends have the same color. We call them *k-colored quadrangulations*, and for the special case when $k = 2$, we will alternatively use the term *bichromatic* quadrangulation. Since monochromatic edges are forbidden in a k-colored quadrangulation, and CH(P) coincides with the outer cycle of *any* quadrangulation of P, CH(P) must be an even-sided *properly colored* polygon, i.e., without monochromatic edges. As in the convex case, we again come up with configurations not admitting k-colored quadrangulations, thus requiring *Steiner points*, that is, auxiliary points which were already mentioned but each point can have any color among the k colors. The bichromatic configuration in the left of Figure 1 is taken from [5].

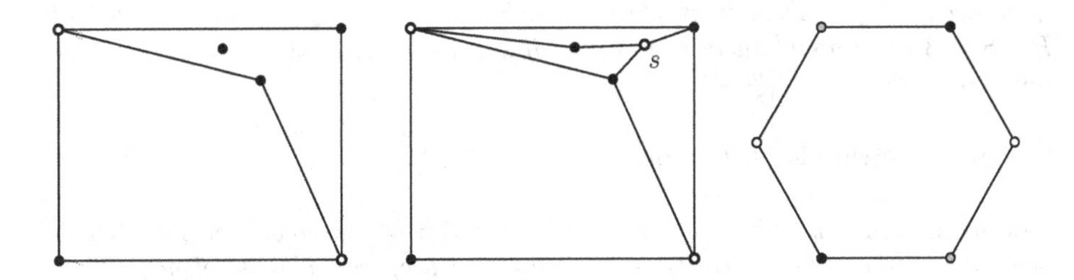

Fig. 1. In the left, a bichromatic point set not admitting a bichromatic quadrangulation. In the middle, the same configuration quadrangulated with a Steiner point s. In the right, a 3-colored point set not admitting a 3-colored quadrangulation no matter how many Steiner points are added.

The study on k-colored quadrangulations of point sets is rather new. One can always construct a bichromatic quadrangulation with the use of roughly $\frac{5n}{12}$ interior Steiner points, and that $\frac{n}{3}$ Steiner points are sometimes necessary [1]. They also considered the case $k = 3$ and showed a surprising fact that there are 3-colored point sets admitting no 3-colored quadrangulations no matter how many interior Steiner points are added, which is definitely an unexpected result. The configuration presented in [1] is shown in the right of Figure 1.

The strange phenomenon for such a 3-colored point set with no 3-colored quadrangulation, even with Steiner points, was recently explained in [7], where the authors showed an elegant characterization of the 3-colored point sets with no 3-colored quadrangulations using Steiner points. The idea was to introduce the notion of "winding number", as explained below:

Let $Q \subset \mathbb{R}^2$ be an m-sided convex polygon on the plane with $m \geq 3$. Let $V(Q)$ denote the vertex set of Q and suppose that Q has a *proper* k-coloring $c : V(Q) \to \{1, \ldots, k\}$, i.e., for any edge xy of Q, $c(x) \neq c(y)$, where $k \geq 2$. Let us define an orientation $\mathcal{O}$ of the edges of Q as follows: if $e = uv$ is an edge of Q, then we orient e from u to v if $c(u) < c(v)$. Let $e^+_{\mathcal{O}}(Q)$ and $e^-_{\mathcal{O}}(Q)$ be the number of edges of Q in clockwise and counter-clockwise direction in $\mathcal{O}$, respectively. The *winding number* of Q, denoted by $\omega(Q)$, is defined as:

$$\omega(Q) = |e^+_{\mathcal{O}}(Q) - e^-_{\mathcal{O}}(Q)|$$

For a k-colored point set P with $k \geq 2$, we define $\omega(P) = \omega(\mathrm{CH}(P))$, extending the definition of winding number for convex polygons to point sets. We say that P can be *k-quadrangulated* or is *k-quadrangulatable* if P admits a k-colored quadrangulation.

The following theorem characterizes the 3-quadrangulatable 3-colored point sets with Steiner points added [7].

Theorem 1 (Kato et al.). *Let $P \subset \mathbb{R}^2$ be a 3-colored n-point set in general position such that* CH(P) *is an m-sided properly colored polygon, where $m \geq 4$ is an even integer. Then there exists a set $S = S(P)$ of Steiner points such that $P \cup S$ is 3-quadrangulatable if and only if $\omega(P) = 0$. If S exists, then S can be taken to be $|S| \leq \frac{7n+17m-48}{18}$.*

Our main contribution is the following result:

Theorem 2. *Let $k \geq 2$ be an integer. Let $P \subset \mathbb{R}^2$ be a k-colored n-point set in general position, where* CH(P) *is an even-sided properly colored polygon. Then there exists a set S of Steiner points such that $P \cup S$ admits a k-quadrangulation if and only if $\omega(P) = 0$ or $k \geq 4$. If S exists, then S can be taken to be $|S| < \frac{(16k-2)n+7k-2}{39k-6}$.*

The condition $\omega(P) = 0$ or $k \geq 4$ in Theorem 2 means that even when only three colors appear on CH(P) and $\omega(P) \neq 0$, we can still find a set S of Steiner points such that $P \cup S$ can be k-quadrangulated as long as we can use at least four colors in total. Moreover, we note that in the case when $k = 2$, $w(P) = 0$ if and only if CH(P) is properly 2-colored.

Our result has the following advantages:

1. Our algorithm for $k = 3$ improves Theorem 1 so much, since our algorithm always perform equally, but the one in Theorem 1 depends on $|\mathrm{CH}(P)|$. (For comparison, our bound for $k = 3$, at worst, is essentially $\frac{46n}{111} < \frac{5n}{12}$, while the one in Theorem 1 can grow larger than n if P has a few interior points.
2. Our result gives the first bounds for the cases when $k \geq 4$.

2 Proof of Theorem 2

In this section, we prove Theorem 2, preparing several lemmas. In those lemmas, we assume that all point sets on the plane are in general position.

The following lemma shown in [7] claims that the winding number is well-defined, and so we may assume that a 3-colored point set is colored by $\{1, 2, 3\}$:

Lemma 1. *Let $Q \subset \mathbb{R}^2$ be a convex polygon with a proper 3-coloring by c_1, c_2, c_3. Then the winding number of Q is invariant for any bijection from $\{c_1, c_2, c_3\}$ to $\{1, 2, 3\}$.*

Lemma 2 is easily obtained from Euler's formula, and used to prove Lemma 3:

Lemma 2. *Let $P \subset \mathbb{R}^2$ be an n-point with $|\mathrm{CH}(P)| = m \geq 4$ even. Then any quadrangulation of P has $(n-1) - \frac{m}{2}$ quadrilaterals and $2(n-1) - \frac{m}{2}$ edges.*

The following two lemmas are main tools for proving the main theorem:

Lemma 3. *Let $Q \subset \mathbb{R}^2$ be a properly m-sided simple convex polygon, where $m \geq 4$ is even. If at least four color appears in Q or $\omega(Q) = 0$, then Q can be partitioned into $\frac{m-2}{2}$ properly colored quadrilaterals.*

Proof. We proceed by induction on m. The case $m = 4$ is trivial, and thus we assume that the lemma holds for every $m' < m$.

CASE 1. *At most three colors appear in Q and $\omega(P) = 0$.*

If P is bichromatic, then the lemma obviously holds, and we may assume that exactly three colors appear in Q. By Lemma 1, suppose that the color classes are $\{1, 2, 3\}$. Observe that there is a vertex $v \in Q$ both of whose neighbors are of the same color. For otherwise, i.e., if every vertex of Q has two neighbors with distinct colors, then we can easily check that Q has a periodic cyclic sequence of colors $1, 2, 3$, which is contrary to $\omega(Q) = 0$. See the left in Figure 2.

Now assume that all edges of Q are oriented by $\mathcal{O}$ explained before. Let $v \in Q$ be a vertex with two neighbors $u, w \in Q$ of the same color, where u and w are the right and the left neighbors of v, respectively. Let $x \in Q$ be the right neighbor

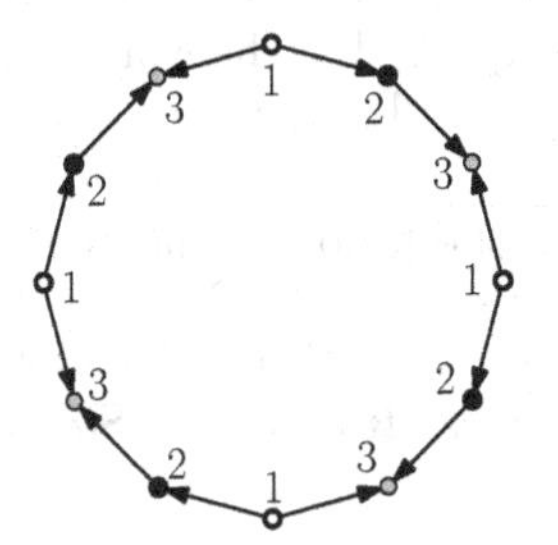

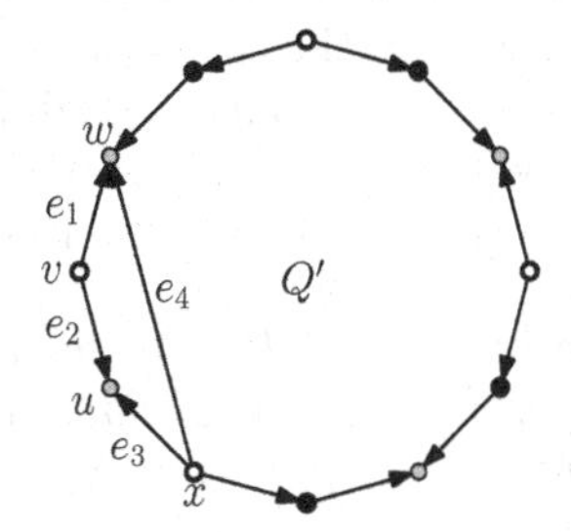

Fig. 2. If Q is colored by the cyclic sequence $1, 2, 3$, as shown in the left, it can be easily verified that $\omega(Q) \neq 0$.

of u. Since Q is properly colored, x has a color distinct from those of u and w, and hence we can add an edge wx to create a properly colored quadrilateral $Q_1 = xuvw$. Now, let Q' be the convex polygon defined by $Q \setminus \{u, v\}$. Observe that $\omega(Q_1) = 0$ since u and w have the same color. Moreover, note that $\omega(Q') = 0$ as well, which is explained as follows: since u and w have the same color, the orientations of vw and vu are canceled in the computation of $\omega(Q)$. Moreover, xu and xw are both oriented away from x or both oriented toward x, so xw is the actual edge making $\omega(Q') = 0$. Hence we get $\omega(Q) = \omega(Q') = 0$. Then we can repeat these procedures inductively on Q', as in the right of Figure 2.

Case 2. *At least four colors appear in Q.*

We claim that there is at least one vertex $w \in Q$ such that at least one of its neighbors at distance 3 on Q is of different color. For otherwise, i.e., if every vertex of Q shares the same color with its two neighbors at distance 3 on Q, then Q would be 3-colored, a contradiction. See the left in Figure 2.

Let $w \in Q$ be one of the vertices having a neighbor at distance 3 of different color, say $x \in Q$. Join w and x with a straight segment, thus creating the quadrilateral, say $Q_1 = wvux$. Let Q' be the convex polygon defined by $Q \setminus \{u, v\}$. If Q' has at least four colors or $\omega(Q') = 0$, then we are done by induction. Otherwise, i.e., if $\omega(Q') \neq 0$ and Q' is 3-colored, then the fourth color appears only at either v or u, say u. In this case, removing the edge wx, we quadrangulate Q by adding diagonals so that all of those are incident to u. Then the resulting quadrangulation must be proper, since no vertex of Q' has the forth color.

In both cases, the total number of created quadrilaterals is $\frac{m-2}{2}$ by Lemma 2.

Lemma 4. *Let $P = C_1 \cup C_2$ be a 2-colored n-point set such that $\mathrm{CH}(P)$ is an m-sided properly colored polygon, where C_1 and C_2 are the color classes of P such that $|C_1| \geq |C_2|$. Then there exists a set $S = S(P)$ of Steiner points such that $P \cup S$ can be 2-quadrangulated, and*

$$|S| \leq \left\lfloor \frac{|C_1|}{3} \right\rfloor + \left\lfloor \frac{|C_2| - \frac{m}{2}}{2} \right\rfloor \leq \frac{5n}{12} - 1.$$

This lemma is essentially one of the main results of [1], and is proven using exactly the same techniques as for Theorem 1 of [1]. However, they are applied differently so the constant term on the bound of $|S|$ is improved in the worst case from $-\frac{1}{3}$ to -1. This negligible improvement of constants will play an important role when proving Theorem 2.

The next lemma is the last one before proceeding to the proof of Theorem 2.

Lemma 5. *Let $P \subset \mathbb{R}^2$ be a k-colored $(q+4)$-point set such that $\mathrm{CH}(P)$ is a properly colored quadrilateral and $k \geq 2$. Then there exist two sets of Steiner points $S_\Gamma = S_\Gamma(P)$ and $S_\Delta = S_\Delta(P)$ such that:*

- *$P \cup S_\Gamma$ can be k-quadrangulated, and $|S_\Gamma| \leq \frac{5q+8}{12}$.*
- *$P \cup S_\Delta$ can be k-quadrangulated, and $|S_\Delta| < \frac{(2k+1)q+16k}{6k}$.*

Proof. We first consider S_Γ. Note that P can be regarded as a bichromatic point set, as follows: if Q is bichromatic itself, say using colors C_1, C_2, then we can recolor every interior point of color different from C_2 with C_1. Rename the color classes as $C_\alpha = C_1$ and $C_\beta = C_2$.

If Q is 3-colored, say using colors C_1, C_2, C_3, then one color, say C_2, appears twice on Q in a diagonal position. Recolor every point of color different from C_2 with a new color C_α, as before. Rename the color class C_2 as C_β.

If Q is 4-colored, say using colors C_1, C_2, C_3, C_4, assume that C_1, C_3 and C_2, C_4 appear in diagonally opposite vertices of Q in clockwise order. Now recolor P with two new colors C_α and C_β as follows: every point of color C_2, C_4 receives color C_β. All other points receive C_α.

Now we have a bichromatic point set with color classes C_α and C_β. Thus by Lemma 4, there exists a set $S_\Gamma = S_\Gamma(P)$ of Steiner points such that $P \cup S_\Gamma$ can be 2-quadrangulated and:

$$|S_\Gamma| \leq \left\lfloor \frac{|C_\alpha|}{3} \right\rfloor + \left\lfloor \frac{|C_\beta| - 2}{2} \right\rfloor \leq \frac{5|P|}{12} - 1 = \frac{5q+8}{12}$$

Second we consider S_Δ. Let $C_1 \cup \cdots \cup C_k$ be the k color classes of *the interior points*, where we let c_i stand for the color of C_i, for $i = 1, \ldots, k$. Suppose that C_1 is a smallest color class, and hence we have $|C_1| \leq \frac{q}{k}$. Suppose that Q is colored with colors other than c_1, since we will later see that this assumption only worsens the upper bound. Now let us introduce two Steiner points x, y of color c_1 inside Q, very close to two opposite vertices of Q, so that the new quadrilateral Q' is still properly colored and contains the q interior points. Let P' be the point set formed by the vertices of Q' and the q points in its interior, see Figure 3.

Now recolor every point of P' with color different from c_1 with a new color C. Then P' can be regarded as a bichromatic point set with color classes $C_1 \cup \{x, y\}$ and C with $(|C_1|+2)+|C|$ vertices, where $|C_1|+|C| = q+2$ and $|C_1| \leq \frac{q}{k}$. We apply Lemma 4 to P', and then the algorithm gives a bichromatic quadrangulation

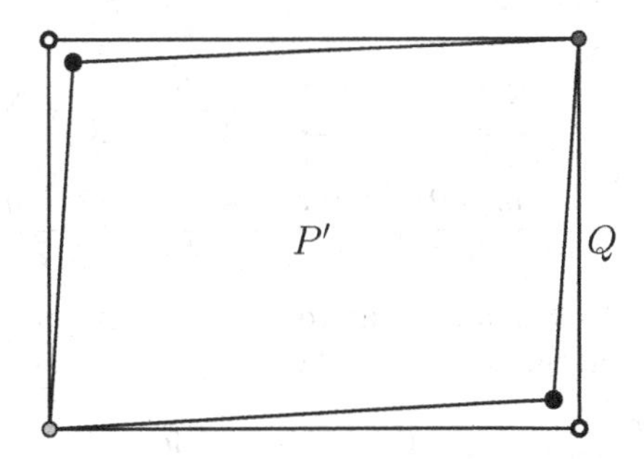

Fig. 3. Points colored with C_1 are represented in black. The quadrilateral Q' still contains the q interior points.

on P with a set of Steiner points $S_\Delta = S_\Delta(P)$ such that:

$$\begin{aligned}
|S_\Delta| &\le \left\lfloor \frac{|C|}{3} \right\rfloor + \left\lfloor \frac{(|C_1|+2)-2}{2} \right\rfloor + 2 \\
&\le \frac{|C|}{3} + \frac{|C_1|}{2} + 2 = \frac{|C|+|C_1|}{3} + \frac{|C_1|}{6} + 2 \\
&\le \frac{q+2}{3} + \frac{q}{6k} + 2 \\
&= \frac{q(2k+1)+16k}{6k}.
\end{aligned}$$

If $|C_1| = \frac{q}{k}$, then $|C_1| = \cdots = |C_k| = \frac{q}{k}$, and hence we can take C_1 so that C_1 appears on Q. In this case, only at most one Steiner point with color c_1 is required to obtain Q'. Hence $|S_\Delta| < \frac{q(2k+1)+7k}{6k}$, since we may assume $|C_1| < \frac{q}{k}$.

We are finally ready to prove Theorem 2:

Proof. We first suppose that $k = 2$. Then we have $\frac{(16k-2)n+7k-2}{39k-6} = \frac{5n+2}{12}$, which is larger than the bound in Lemma 4. Hence the theorem follows for $k = 2$, and so we may assume $k \ge 3$.

Let P be a k-colored n-point set, where let $|\mathrm{CH}(P)| = m$ and $q = n - m$ be the number of interior points. Let k' be the number of colors appearing on $\mathrm{CH}(P)$. If $k' \ge 4$ or $\omega(P) = 0$, then by Lemma 3, we can partition $\mathrm{CH}(P)$ into $r = \frac{m-2}{2}$ properly colored convex quadrilaterals Q_i, for $i = 1, \ldots r$. However, we might have $k' = 3$ (but $k \ge 4$) and $\omega(P) \ne 0$. In this case, we put exactly one inner Steiner point, say s, with a color not appearing in $\mathrm{CH}(P)$, very closed to one of the points in $\mathrm{CH}(P)$, say v. If we put $P' = (\mathrm{CH}(P) - \{v\}) \cup \{s\}$, then we can apply Lemma 3 to P', since P' has at least four colors in $\mathrm{CH}(P')$. Clearly, if the theorem holds for P', then it also holds for P but we need one more Steiner point corresponding to s.

Let q_i be the number of interior points in the quadrilateral Q_i, for $i = 1, \ldots, r$. By Lemma 5, there are two ways to k-quadrangulate Q_i using Steiner points. The first way uses a set of Steiner points S_Γ^i for each Q_i, which gives a k-quadrangulation of $P \cup S_\Gamma$ with $S_\Gamma = S_\Gamma^1 \cup \cdots \cup S_\Gamma^r$:

$$\begin{aligned}
|S_\Gamma| &= \sum_{i=1}^{r} |S_\Gamma^i| \le \sum_{i=1}^{r} \frac{5q_i+8}{12} = \frac{2r}{3} + \sum_{i=1}^{r} \frac{5q_i}{12} \\
&= \frac{m-2}{3} + \frac{5q}{12} = \frac{m-2}{3} + \frac{5(n-m)}{12} \qquad (1)
\end{aligned}$$

The second way to k-quadrangulate each Q_i using a set S_Δ^i of Steiner points gives a k-quadrangulation of $P \cup S_\Delta$ with $S_\Delta = S_\Delta^1 \cup \cdots \cup S_\Delta^r$ such that:

$$\begin{aligned}
|S_\Delta| &= \sum_{i=1}^{r} |S_\Delta^i| < \sum_{i=1}^{r} \frac{(2k+1)q_i+16k}{6k} = \frac{8r}{3} + \sum_{i=1}^{r} \frac{(2k+1)q_i}{6k} \\
&= \frac{4(m-2)}{3} + \frac{(2k+1)q}{6k} = \frac{4(m-2)}{3} + \frac{(2k+1)(n-m)}{6k} \qquad (2)
\end{aligned}$$

Using (1) and (2), we consider which of S_Γ or S_Δ performs better:

$$\frac{m-2}{3}+\frac{5(n-m)}{12}<\frac{4(m-2)}{3}+\frac{(2k+1)(n-m)}{6k} \iff m>\frac{k(n+24)-2n}{13k-2} \quad (3)$$

For a given n-point set P, we construct a set S of Steiner points such that $P \cup S$ admits a k-quadrangulation. Hence, if P satisfies $m > \frac{k(n+24)-2n}{13k-2}$, then S_Γ performs better by (3). So we let $S = S_\Gamma$ or $S = S_\Gamma \cup \{s\}$, and hence

$$|S| \leq |S_\Gamma|+1 \leq \frac{m-2}{3}+\frac{5(n-m)}{12}+1<\frac{(16k-2)n+7k-2}{39k-6}$$

On the other hand, if $m \leq \frac{k(n+24)-2n}{13k-2}$, then S_Δ performs better, and hence let $S = S_\Delta$ or $S = S_\Delta \cup \{s\}$, and

$$|S| \leq |S_\Delta|+1<\frac{4(m-2)}{3}+\frac{(2k+1)(n-m)}{6k}+1 \leq \frac{(16k-2)n+7k-2}{39k-6}$$

Now Theorem 2 follows entirely.

3 *k*-Colored Sets of Points Requiring Many Steiner Points

It was shown in [1] that there are bicolored sets P of $n = 3m$ points with $m \geq 4$ that requires at least m Steiner points to be 2-quadrangulated, where $m = |\mathrm{CH}(P)|$. See the left in Figure 4.

We can briefly describe the configuration P presented in [1]: For each edge e of $\mathrm{CH}(P)$, exactly two interior points p_e, q_e are associated, as shown in the left of Figure 4. The coloring of the configuration is done in the following way: the endpoints of e get different colors and its associated pair of interior points gets the color of the left endpoint of e, as in the left of Figure 4. Let G be any

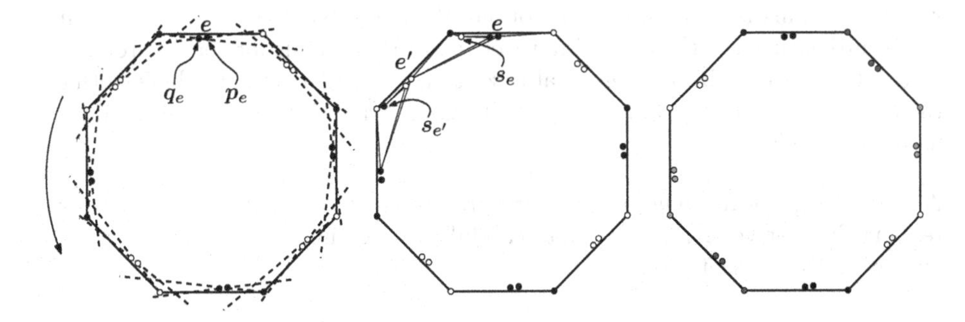

Fig. 4. In the left the bichromatic configuration P that needs at least $\frac{n}{3}$ Steiner points in order to be 2-quadrangulated. Every edge e of $\mathrm{CH}(P)$ gets associated with a pair of interior points p_e, q_e. In the middle a partial bichromatic quadrangulation using Steiner points $s_e \neq s_{e'}$ is shown. In the right the same configuration colored with 4 colors.

2-colored quadrangulation on $P \cup S$ with Steiner points S. Then, if we let f_e be the quadrilateral face of G incident to e, then f_e must require at least one Steiner point, say s_e, since p_e, q_e and the left point of e are monochromatic. Intuitively, s_e locally helps to k-quadrangulate the region between e and p_e, q_e, see the middle of Figure 4.

For the k-colored case, the number of case analysis increases, since there are more colors for the Steiner points s_e. The reader will be able to verify that the same arguments as in [1] carry over into the k-colored setting, and hence we also get the lower bound of $\frac{n}{3}$ Steiner points. We will thus refrain ourselves to repeating those arguments here. See the right in Figure 4 for an example of a 4-colored configuration needing at least $\frac{n}{3}$ Steiner points. Hence we have:

Proposition 1. *For any $k \geq 2$, there exists a k-colored n-point set P satisfying the following: if S is any set of Steiner points such that $P \cup S$ admits a k-colored quadrangulation, then $|S| \geq \frac{n}{3}$.*

4 Conclusions

In this paper, we have studied the problem of constructing k-colored quadrangulations of k-colored sets of points using Steiner points, for any $k \geq 2$. Moreover, we have been able to improve the previous known upper bound for the number of Steiner points when $k = 3$, and given the first upper bounds for the case when $k \geq 4$. We also pointed out that the lower bound of $\frac{n}{3}$ interior Steiner points for the bichromatic case [1] follows in more general cases $k \geq 3$.

The upper bound for the number of Steiner points in the k-colored case is always less than $\frac{(16k-2)n+7k-2}{39k-6}$, in which we essentially have $\frac{16n}{39} \approx 0.4102n$. Since $\frac{n}{3}$ is the lower bound, both bounds differ by roughly $\frac{n}{13}$. Therefore closing this gap is still an interesting open question.

There is one more thing to note. The upper bound in the bichromatic case is roughly $\frac{5n}{12} = 0.41\bar{6}n$ [1]. Thus both upper bounds are essentially the same in the worst case. This is because the core of our algorithm bases on the technique in the bichromatic case. Hence, an improvement of the bounds for the bichromatic case will carry over into the general case using our algorithm. We believe that the cases $k = 2$ and $k = 3$ are really challenging, while the cases $k \geq 4$ might be more attackable.

Acknowledgement. The authors are grateful to the anonymous referee for reading the first version of the paper carefully and giving us several suggestions improving the presentation.

References

1. Alvarez, V., Sakai, T., Urrutia, J.: Bichromatic quadrangulations with steiner points. Graph. Combin. 23, 85–98 (2007)
2. Bremner, D., Hurtado, F., Ramaswami, S., Sacristan, V.: Small strictly convex quadrilateral meshes of point sets. Algorithmica 38, 317–339 (2003)

3. Bose, P., Toussaint, G.T.: Characterizing and efficiently computing quadrangulations of planar point sets. Computer Aided Geometric Design 14, 763–785 (1997)
4. Bose, P., Ramaswami, S., Toussaint, G.T., Turki, A.: Experimental results on quadrangulations of sets of fixed points. Computer Aided Geometric Design 19, 533–552 (2007)
5. Cortés, C., Márquez, A., Nakamoto, A., Valenzuela, J.: Quadrangulations and 2-colorations. In: EuroCG, pp. 65–68. Technische Universiteit Eindhoven (2005)
6. Heredia, V.M., Urrutia, J.: On convex quadrangulations of point sets on the plane. In: Akiyama, J., Chen, W.Y.C., Kano, M., Li, X., Yu, Q. (eds.) CJCDGCGT 2005. LNCS, vol. 4381, pp. 38–46. Springer, Heidelberg (2007)
7. Kato, S., Mori, R., Nakamoto, A.: Quadrangulations on 3-colored point sets with steiner points and their winding number. Graphs Combin., DOI 10.1007/s00373-013-1346-4
8. Lai, M.-J., Schumaker, L.L.: Scattered data interpolation using c2 supersplines of degree six. SIAM J. Numer. Anal. 34, 905–921 (1997)
9. Ramaswami, S., Ramos, P.A., Toussaint, G.T.: Converting triangulations to quadrangulations. Comput. Geom. 9, 257–276 (1998)
10. Schiffer, T., Aurenhammer, F., Demuth, M.: Computing convex quadrangulations. Discrete Applied Math. 160, 648–656 (2012)
11. Toussaint, G.T.: Quadrangulations of planar sets. In: Sack, J.-R., Akl, S.G., Dehne, F., Santoro, N. (eds.) WADS 1995. LNCS, vol. 955, pp. 218–227. Springer, Heidelberg (1995)

On Universal Point Sets for Planar Graphs*

Jean Cardinal[1], Michael Hoffmann[2], and Vincent Kusters[2]

[1] Département d'Informatique
Université Libre de Bruxelles (ULB)
jcardin@ulb.ac.be
[2] Institute of Theoretical Computer Science
ETH Zürich
{hoffmann,vincent.kusters}@inf.ethz.ch

Abstract. A set P of points in $\mathbb{R}^2$ is n-universal, if every planar graph on n vertices admits a plane straight-line embedding on P. Answering a question by Kobourov, we show that there is no n-universal point set of size n, for any $n \geq 15$. Conversely, we use a computer program to show that there exist universal point sets for all $n \leq 10$ and to enumerate all corresponding order types. Finally, we describe a collection $\mathcal{G}$ of $7'393$ planar graphs on 35 vertices that do not admit a simultaneous geometric embedding without mapping, that is, no set of 35 points in the plane supports a plane straight-line embedding of all graphs in $\mathcal{G}$.

1 Introduction

We consider plane, straight-line embeddings of graphs in $\mathbb{R}^2$. In those embeddings, vertices are represented by pairwise distinct points, every edge is represented by a line segment connecting its endpoints, and no two edges intersect except at a common endpoint.

An *n-universal* (or short *universal*) point set for planar graphs admits a plane straight-line embedding of all graphs on n vertices. A longstanding open problem is to give precise bounds on the minimum number of points in an n-universal point set. The currently known asymptotic bounds are apart by a linear factor. On the one hand, it is known that every planar graph can be embedded on a grid of size $n-1 \times n-1$ [1,2]. On the other hand, it was shown that at least $1.235n$ points are necessary [3], improving earlier bounds of $1.206n$ [4] and $n + \sqrt{n}$ [1].

The following, somewhat simpler question was asked ten years ago by Kobourov [5]: what is the largest value of n for which a universal point set of size n exists? We prove the following.

Theorem 1. *There is no n-universal point set of size n, for any $n \geq 15$.*

* The research was carried out while the first author was at ETH Zürich. Research of J. Cardinal is partially supported by the ESF EUROCORES programme EuroGIGA, CRP ComPoSe. Research of M. Hoffmann and V. Kusters is partially supported by the ESF EUROCORES programme EuroGIGA, CRP GraDR and the Swiss National Science Foundation, SNF Project 20GG21-134306.

J. Akiyama, M. Kano, and T. Sakai (Eds.): TJJCCGG 2012, LNCS 8296, pp. 30–41, 2013.

At some point, the Open Problem Project page dedicated to the problem [5] mentioned that Kobourov proved there exist 14-universal point sets of size 14. If this is correct, our bound is tight, and the answer to the above question is $n = 14$. After verification, however, this claim appears to be unsubstantiated [6]. We managed to check that there exist universal point sets only up to $n \leq 10$. Further investigations are ongoing.

Overview. Section 2 is devoted to the proof of Theorem 1. It combines a labeled counting scheme for *planar 3-trees* (also known as *stacked triangulations*) that is very similar to the one used by Kurowski in his asymptotic lower bound argument [3] with known lower bounds on the rectilinear crossing number [7,8]. Note that although planar 3-trees seem to be useful for lower bounds, a recent preprint from Fulek and Tóth [9] shows that there exist n-universal point sets of size $O(n^{5/3})$ for planar 3-trees.

For a collection $\mathcal{G} = \{G_1, \ldots, G_k\}$ of planar graphs on n vertices, a *simultaneous geometric embedding without mapping* for $\mathcal{G}$ is a collection of plane straight-line embeddings $\phi_i : G_i \to P$ onto the same set $P \subset \mathbb{R}^2$ of n points.

In Section 3, we consider the following problem: what is the largest natural number σ such that every collection of σ planar graphs on the same number of vertices admit a simultaneous geometric embedding without mapping? From the Fáry-Wagner Theorem [10,11] we know that $\sigma \geq 1$. We prove the following upper bound:

Theorem 2. *There is a collection of* 7′393 *planar graphs on* 35 *vertices that do not admit a simultaneous plane straight-line embedding without mapping, hence* $\sigma < 7'393$.

To our knowledge these are the best bounds currently known. It is a very interesting and probably challenging open problem to determine the exact value of σ.

Finally, in Section 4, we use a computer program to show that there exist n-universal point sets of size n for all $n \leq 10$ and give the total number of such point sets for each n. As a side remark, note that it is not clear that the property "there exists an n-universal point set of size n" is monotone in n.

2 Large Universal Point Sets

A planar 3-tree is a maximal planar graph obtained by iteratively splitting a facial triangle into three new triangles with a degree-three vertex, starting from a single triangle. Since a planar 3-tree is a maximal planar graph, it has n vertices and $2n - 4$ triangular faces and its combinatorial embedding is fixed up to the choice of the outer face.

For every integer $n \geq 4$, we define a family $\mathcal{T}_n$ of labeled planar 3-trees on the set of vertices $[n] := \{1, \ldots, n\}$ as follows:

(i) $\mathcal{T}_4$ contains only the complete graph K_4,

(ii) $\mathcal{T}_n$ contains every graph that can be constructed by making the new vertex n adjacent to the three vertices of one of the $2n - 6$ facial triangles of some $T \in \mathcal{T}_{n-1}$.

We insist on the fact that $\mathcal{T}_n$ is a set of *labeled* abstract graphs, many of which can in fact be isomorphic if considered as abstract (unlabeled) graphs. We also point out that for $n > 4$, the class $\mathcal{T}_n$ does not contain *all* labeled planar 3-trees on n vertices. For instance, the four graphs in $\mathcal{T}_5$ are shown in Fig. 1, and there is no graph for which both Vertex 1 and Vertex 2 have degree three.

Lemma 1. *For $n \geq 4$, we have $|\mathcal{T}_n| = 2^{n-4} \cdot (n-3)!$.*

Proof. By definition, $|\mathcal{T}_4| = 1$. Every graph in $\mathcal{T}_n$ is constructed by splitting one of the $2n - 6$ faces of a graph in $\mathcal{T}_{n-1}$. We therefore have:

$$|\mathcal{T}_n| = |\mathcal{T}_{n-1}| \cdot (2n-6) = 4 \cdot 6 \cdot \ldots \cdot (2n-6) = 2^{n-4} \cdot (n-3)!. \qquad \square$$

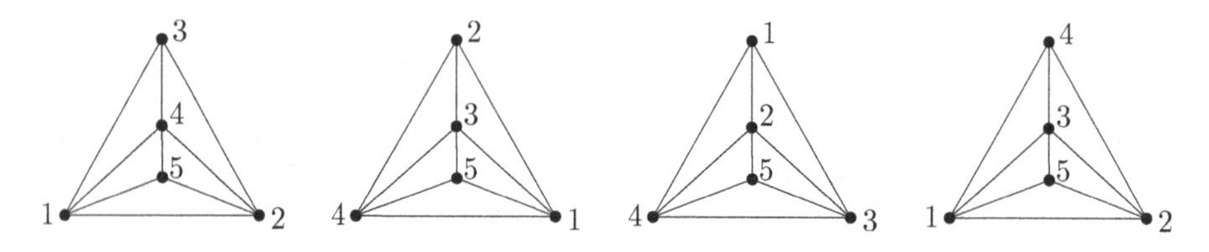

Fig. 1. The four planar 3-trees in $\mathcal{T}_5$, with vertex set $\{1, 2, 3, 4, 5\}$

Lemma 2. *Given a set $P = \{p_1, \ldots, p_n\}$ of labeled points in the plane and a bijection $\pi : [n] \rightarrow P$, there is* at most one *$T \in \mathcal{T}_n$ such that π is a plane straight-line embedding of T.*

Proof. Consider any such labeled point set P and assume without loss of generality that $\pi(i) = p_i$ for all i. In all $T \in \mathcal{T}_n$ the vertices $\{1, 2, 3, 4\}$ form a K_4. Hence, for all T, the straight-line embedding π connects all pairs of points in $\{p_1, p_2, p_3, p_4\}$ with line segments. If these points are in convex position, there is a crossing and there is no $T \in \mathcal{T}_n$ for which π is a plane straight-line embedding (Fig. 2). Otherwise, there is a unique graph $K_4 \in \mathcal{T}_4$ for which p_1, p_2, p_3, p_4 is a plane straight-line drawing. We proceed as follows.

Given a plane straight-line drawing for some graph $T_i \in \mathcal{T}_i$ on the first $i \geq 4$ points, the next point p_{i+1} is located in some triangular region of the drawing; denote this region by $p_{a_i} p_{b_i} p_{c_i}$. Only if during the construction of T we decided to connect the next vertex $i + 1$ to exactly the vertices a_i, b_i, c_i, there is no crossing introduced by mapping $i + 1$ to p_{i+1}. (An edge to any other vertex would cross one of the bounding edges of the triangle $p_{a_i} p_{b_i} p_{c_i}$.) In other words, for every $i \geq 5$ the role of vertex i is completely determined. If no crossing is ever introduced, this process determines exactly one graph $T \in \mathcal{T}_n$ for which π forms a plane straight-line embedding. (Note that a crossing can be introduced only if p_{i+1} is located outside of the convex hull of $\{p_1, \ldots, p_i\}$. And also in that case there need not be a crossing, as the example in Fig. 2 (right) shows.) $\square$

Fig. 2. Some permutations of a given point set do not define any planar 3-tree in $\mathcal{T}_n$, because they generate a crossing (left). On the other hand, when no such crossing occurs, the permutation defines a unique planar 3-tree in $\mathcal{T}_n$ (right). At any rate, a single permutation can be associated with at most one planar 3-tree in $\mathcal{T}_n$.

We use the following theorem by Ábrego and Fernández-Merchant.

Theorem 3 ([7]). *Every plane straight-line drawing of the complete graph K_n has at least $\frac{1}{4}\left\lfloor\frac{n}{2}\right\rfloor\left\lfloor\frac{n-1}{2}\right\rfloor\left\lfloor\frac{n-2}{2}\right\rfloor\left\lfloor\frac{n-3}{2}\right\rfloor$ crossings.*

Note that for $n \leq 4$ at least one of the floor expressions is zero, whereas for $n = 5$ the theorem states that every straight-line drawing of K_5 has at least one crossing. Any pair of crossing edges corresponds to a four-tuple of points in convex position. Using this interpretation we can easily derive a floor-free lower bound on the number of convex four-gons contained in every planar point set.

Corollary 1. *Given a point set $P \subset \mathbb{R}^2$ of n points in general position, more than a $\frac{3}{8} \cdot \frac{n-4}{n}$-fraction of all four element subsets of P is in convex position.*

Proof. By Theorem 3 at least $c = \frac{1}{4}\left\lfloor\frac{n}{2}\right\rfloor\left\lfloor\frac{n-1}{2}\right\rfloor\left\lfloor\frac{n-2}{2}\right\rfloor\left\lfloor\frac{n-3}{2}\right\rfloor$ four element subsets of P are in convex position. For n odd we have

$$c = \frac{1}{4}\left(\frac{n-1}{2}\right)^2\left(\frac{n-3}{2}\right)^2$$

and for n even we have

$$c = \frac{1}{4}\left(\frac{n}{2}\right)\left(\frac{n-2}{2}\right)^2\left(\frac{n-4}{2}\right)$$

and so

$$\begin{aligned} c &> \frac{1}{4}\left(\frac{n-1}{2}\right)\left(\frac{n-2}{2}\right)\left(\frac{n-3}{2}\right)\left(\frac{n-4}{2}\right) \\ &= \frac{3}{8} \cdot \frac{n-4}{n} \cdot \frac{n(n-1)(n-2)(n-3)}{4 \cdot 3 \cdot 2} \\ &= \frac{3}{8} \cdot \frac{n-4}{n} \cdot \binom{n}{4}, \end{aligned}$$

for all n. □

We will use this fact to prove the following lemma.

Lemma 3. *On any set $P \subset \mathbb{R}^2$ of $n \geq 4$ points fewer than $\frac{1}{8}(5n+12)(n-1)!$ graphs from $\mathcal{T}_n$ admit a plane straight-line embedding.*

Proof. Let $P \subset \mathbb{R}^2$ be a set of n points and denote by $\mathcal{F}_n \subseteq \mathcal{T}_n$ the set of labeled planar 3-trees from $\mathcal{T}_n$ that admit a plane straight-line embedding onto P. Note that a straight-line embedding can be represented by a permutation π of the points of P, where each vertex i is mapped to point $\pi(i)$. Let S_n be the set of all permutations of P. We define a map $\psi : \mathcal{F}_n \to S_n$ by assigning to each $T \in \mathcal{F}_n$ some $\psi(T) \in S_n$ such that $\psi(T)$ is a plane straight-line embedding of T (such an embedding exists by definition of $\mathcal{F}_n$).

By Lemma 2, every permutation $\pi \in S_n$ is a plane straight-line embedding of *at most one* $T \in \mathcal{F}_n$. It follows that ψ is a injection, and hence $\psi : \mathcal{F}_n \to \Pi$, with $\Pi = \text{Im}(\psi)$, is a bijection and so $|\mathcal{F}_n| = |\Pi| \leq |S_n| = n!$.

Next we can quantify the difference between Π and S_n using Corollary 1. Note that the general position assumption is not a restriction, since in case of collinearities, a slight perturbation of the point set yields a new point set that still admits all plane straight-line drawings of the original point set. Consider a permutation $\pi = p_1, \ldots, p_n$ such that p_1, p_2, p_3, p_4 form a convex quadrilateral. As argued in the first paragraph of the proof of Lemma 2, π is not a plane straight-line embedding for any $T \in \mathcal{F}_n$. It follows that $\pi \in S_n \backslash \Pi$. We know from Corollary 1 that more than a fraction of $(3/8)\cdot(n-4)/n$ of the 4-tuples of P are in convex position and therefore a corresponding fraction of all permutations does *not* correspond to a plane straight-line drawing. So we can bound the number of possible labeled plane straight-line drawings by

$$|\Pi| < \left(1 - \frac{3}{8} \cdot \frac{n-4}{n}\right) n! = \frac{1}{8}(5n+12)(n-1)!\,. \qquad \square$$

Proof (of Theorem 1). Consider an n-universal point set $P \subset \mathbb{R}^2$ with $|P| = n$. Being universal, in particular P has to accommodate all graphs from $\mathcal{T}_n$. By Lemma 1, there are exactly $2^{n-4} \cdot (n-3)!$ graphs in $\mathcal{T}_n$, whereas by Lemma 3 no more than $\frac{1}{8}(5n+12)(n-1)!$ graphs from $\mathcal{T}_n$ admit a plane straight-line drawing on P. Combining both bounds we obtain $2^{n-1} \leq (5n+12)(n-1)(n-2)$. Setting $n = 15$ yields $2^{14} = 16'384 \leq 87 \cdot 14 \cdot 13 = 15'834$, which is a contradiction and so there is no 15-universal set of 15 points.

For $n = 14$ the inequality reads $2^{13} = 8'192 \leq 82 \cdot 13 \cdot 12 = 12'792$ and so there is no indication that there cannot be a 14-universal set of 14 points. To prove the claim for any $n > 15$, consider the two functions $f(n) = 2^{n-1}$ and $g(n) = (5n+12)(n-1)(n-2)$ that constitute the inequality. As f is exponential in n whereas g is just a cubic polynomial, f certainly dominates g, for sufficiently large n. Moreover, we know that $f(15) > g(15)$. Noting that $f(n)/f(n-1) = 2$ and $g(n) > 0$, for $n > 2$, it suffices to show that $g(n)/g(n-1) < 2$, for all $n \geq 16$.

We can bound

$$\frac{g(n)}{g(n-1)} = \frac{(5n+12)(n-1)(n-2)}{(5(n-1)+12)(n-2)(n-3)} = \frac{(5n+12)(n-1)}{(5n+7)(n-3)}$$
$$< \frac{(5n+15)n}{5n(n-3)} = \frac{n+3}{n-3},$$

which is easily seen to be upper bounded by two, for $n \geq 9$. □

3 Simultaneous Geometric Embeddings

The number of non-isomorphic planar 3-trees on n vertices was computed by Beineke and Pippert [12], and appears as sequence A027610 on Sloane's Encyclopedia of Integer Sequences. For $n = 15$, this number is $321'776$. Hence we can also phrase our result in the language of simultaneous embeddings [13].

Corollary 2. *There is a collection of* $321'776$ *planar graphs that do not admit a simultaneous (plane straight-line) embedding without mapping.*

In the following we will give an explicit construction for a much smaller family of graphs that not admit a simultaneous embedding without mapping. As a first observation, note that the freedom to select the outer face is essential in order to embed graphs onto a given point set. In fact, for planar 3-trees, the mapping for the outer face is the only choice there is. We prove this in two steps.

Lemma 4. *Let G be a labeled planar 3-tree on the vertex set $[n]$, for $n \geq 3$, and let C denote any triangle in G. Then G can be constructed starting from C by iteratively inserting a degree-three vertex into some facial triangle of the partial graph constructed so far.*

Proof. We prove the statement by induction on n. For $n = 3$ there is nothing to show. Hence let $n > 3$. By definition G can be constructed iteratively from *some* triangle in the way described. Without loss of generality suppose that adding vertices in the order $1, 2, \ldots, n$ yields such a construction sequence. Denote by G_i the graph that is constructed by the sequence $1, \ldots, i$, for $1 \leq i \leq n$.

Let $C = u, v, w$ such that $u < v < w$. Consider the graph G_w: In the last step, w is added as a new vertex into some facial triangle T of G_{w-1}. As w is a neighbor of both u and v in G, both u and v are vertices of T; denote the third vertex of T by x. Note that all of u, v, w and u, w, x and v, x, w are facial triangles in G_w.

If $w = 4$, then exchanging the role of w and x yields a construction sequence $u, v, w, x, 5, \ldots, n$ for G, as claimed. If $w > 4$, then c_1, c_2, v is a separating triangle in G_w. By the inductive hypothesis we can obtain a construction sequence S for G_{w-1} starting with the triangle u, v, x. The desired sequence for G is obtained as $u, v, w, x, S^-, w+1, \ldots, n$, where S^- is the suffix of S that excludes the starting triangle u, v, x. □

And now we can prove the desired property:

Lemma 5. *Given a labeled planar 3-tree G on vertex set $[n]$, a triangle $c = c_1c_2c_3$ in G, and a set $P \subset \mathbb{R}^2$ of n points with $p_1, p_2, p_3 \in P$, there is at most one way to complete the partial embedding $\{c_1 \mapsto p_1, c_2 \mapsto p_2, c_3 \mapsto p_3\}$ to a plane straight-line embedding of G on P.*

Proof. We use Lemma 4 to relabel the vertices in such a way that c_1, c_2, c_3 becomes $1, 2, 3$ and the order $1, \ldots, n$ is a construction sequence for G. Embed vertices $1, 2, 3$ onto p_1, p_2, p_3. We iteratively embed the remaining vertices as follows. Vertex i was inserted into some face $jk\ell$ during the construction given by Lemma 4. Note that j, k, ℓ have already been embedded on points p_j, p_k, p_ℓ. The vertices contained in the triangle $jk\ell$ (except i) are partitioned into three sets by the cycles ijk (n_1 vertices) and $ik\ell$ (n_2 vertices) and $i\ell j$ (n_3 vertices). We want to embed i on a point p_i such that $p_ip_jp_k$ contains exactly n_1 points, $p_ip_kp_\ell$ contains exactly n_2 points and $p_ip_\ell p_j$ contains exactly n_3 points. Note that it is necessary to embed i on a point with this property: if some triangle has too few points, then it will not be possible to embed the subgraph of G enclosed by the corresponding cycle there. It remains to show that there is always at most one choice for p_i. Suppose that there are two candidates for p_i, say p_i' and p_i''. Then p_i'' must be contained in $p_i'p_jp_k$ or $p_i'p_kp_\ell$ or $p_i'p_\ell p_j$ (or vice versa). Without loss of generality, let it be contained in $p_i'p_jp_k$: now $p_i''p_jp_k$ contains fewer points than $p_i'p_jp_k$, which is a contradiction. The lemma follows by induction. □

Therefore it is not surprising that it is very easy to find three graphs that do not admit a simultaneous (plane straight-line) embedding without mapping, if the mapping for the outer face is specified for each of them.

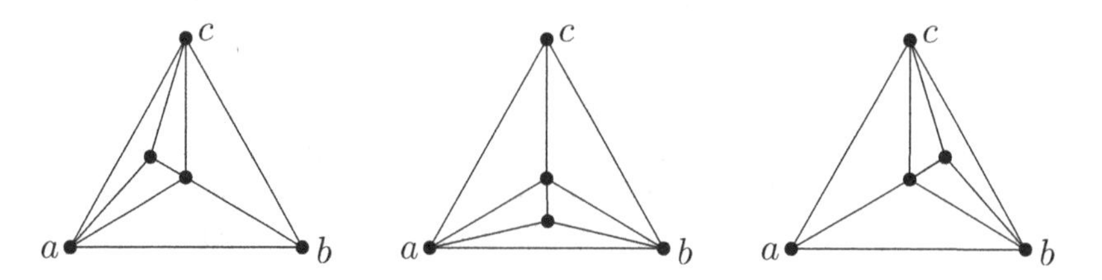

Fig. 3. Three planar graphs that do not admit a simultaneous geometric embedding with a fixed mapping for the outer face

Lemma 6. *There is no set $P \subset \mathbb{R}^2$ of five points with convex hull p_a, p_b, p_c such that every graph shown in Fig. 3 has a (plane straight-line) embedding on P where the vertices a, b and c are mapped to the points p_a, p_b and p_c, respectively.*

Proof. The point p for the central vertex that is connected to all of a, b, c must be chosen so that (i) it is not in convex position with p_a, p_b and p_c and (ii) the number of points in the three resulting triangles is one in one triangle and zero in the other two. That requires three distinct choices for p, but there are only two points available. □

In fact, there are many such triples of graphs. The following lemma can be verified with help of a computer program that exhaustively checks all order types. Point set order types [14] are a combinatorial abstraction of planar point sets that encode the orientation of all point triples, which in particular determines whether or not any two line segments cross. For a small number of points, there is a database with realizations of every (realizable) order type [15].

Lemma 7. *There is no set $P \subset \mathbb{R}^2$ of eight points with convex hull p_a, p_b, p_c such that every graph shown in Fig. 4 has a (plane straight-line) embedding on P where the vertices a, b and c are mapped to the points p_a, p_b and p_c, respectively.*

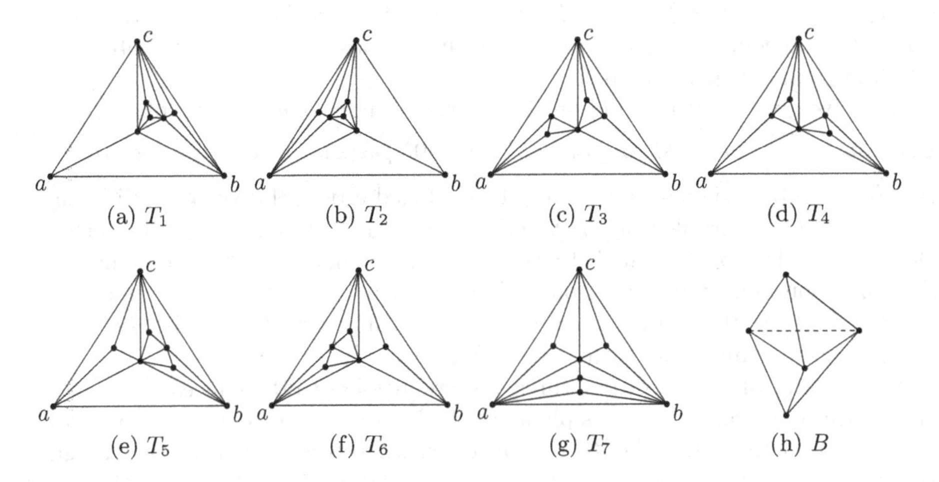

Fig. 4. (a)–(g): Seven planar graphs, no three of which admit a simultaneous geometric embedding with a fixed mapping for the outer face; (h): the skeleton B of a triangular bipyramid

Denote by $\mathcal{T} = \{T_1, \ldots, T_7\}$ the family of seven graphs on eight vertices depicted in Fig. 4. We consider these graphs as abstract but *rooted* graphs, that is, one face is designated as the outer face and the counterclockwise order of the vertices along the outer face (the *orientation* of the face) is a, b, c in each case. Observe that all graphs in $\mathcal{T}$ are planar 3-trees.

Using $\mathcal{T}$ we construct a family $\mathcal{G}$ of graphs as follows. Start from the skeleton B of a triangular bipyramid, that is, a triangle and two additional vertices, each of which is connected to all vertices of the triangle. The graph B has five vertices and six faces and it is a planar 3-tree.

We obtain $\mathcal{G}$ from B by planting one of the graphs from $\mathcal{T}$ onto each of the six faces of B. Each face of B is a (combinatorial) triangle where one vertex has degree three (one of the pyramid tips) and the other two vertices have degree four (the vertices of the starting triangle). On each face f of B a selected graph T from $\mathcal{T}$ is planted by identifying the three vertices bounding f with the three vertices bounding the outer face of T in such a way that vertex c (which appears at the top in Fig. 4) is mapped to the vertex of degree three (in B) of f. In the next paragraph, we will see why we do not have to specify how a and b are matched to f. The family $\mathcal{G}$ consists of all graphs on $5 + 6 \cdot 5 = 35$ vertices that can be obtained in this way. By construction all these graphs are planar 3-trees. Therefore by Lemma 5 on any given set of 35 points, the plane straight-line embedding is unique (if it exists), once the mapping for the outer face is determined.

Observe that $\mathcal{T}$ is *flip-symmetric* with respect to horizontal reflection. In other (more combinatorial) words, for every $T \in \mathcal{T}$ we can exchange the role of the bottom two vertices a and b of the outer face (and thereby also its orientation) to obtain a graph that is also in $\mathcal{T}$. The graphs form symmetric pairs of siblings (T_1, T_2), (T_3, T_4), (T_5, T_6), and T_7 flips to itself. Therefore, regardless of the orientation in which we plant a graph from $\mathcal{T}$ onto a face of B, we obtain a graph in $\mathcal{G}$, and so $\mathcal{G}$ is well-defined.

Next, we give a lower bound on the number of nonisomorphic graphs in $\mathcal{G}$.

Lemma 8. *The family $\mathcal{G}$ contains at least* $9'805$ *pairwise nonisomorphic graphs.*

Proof. Consider the bipyramid B as a face-labeled object. There are 7^6 different ways to assign a graph from $\mathcal{T}$ to each of the six now distinguishable faces. Denote this class of face-labeled graphs by $\mathcal{F}$. For many of these assignments the corresponding graphs are isomorphic if considered as abstract (unlabeled) graphs. However, the following argument shows that every isomorphism between two such graphs maps the vertex set of B to itself.

The two tips of B have degree three and are incident to three faces. Onto each of the faces one graph from $\mathcal{T}$ is planted, which increases the degree by four (for $T_1, \ldots, T_6$) or three (for T_7) to a total of at least twelve. The three triangle vertices start with degree four and are incident to four faces. Every graph from $\mathcal{T}$ planted there adds at least one more edge, to a total degree of at least eight. But the highest degree among the interior vertices of the graphs in $\mathcal{T}$ is seven, which proves the claim.

Hence we have to look for isomorphisms only among the symmetries of the bipyramid B. The tips are distinguishable from the triangle vertices, because the former are incident to three high degree vertices, whereas the latter are incident to four high degree vertices. Selecting the mapping for one face of B determines the whole isomorphism. Since there are at most two ways to map a face to a face (we can select the mapping for the two non-tip vertices, that is, the orientation of the triangle), every graph in $\mathcal{F}$ is isomorphic to at most $2 \cdot 6 = 12$ graphs from $\mathcal{F}$. It follows that there are at least $7^6/12 > 9'804$ pairwise nonisomorphic graphs in $\mathcal{G}$. □

We now give an upper bound on the number of graphs of $\mathcal{G}$ that can be simultaneously embedded on a common point set.

Lemma 9. *At most* $7'392$ *pairwise nonisomorphic graphs of $\mathcal{G}$ admit a simultaneous (plane straight-line) embedding without mapping.*

Proof. Consider a subset $\mathcal{G}' \subseteq \mathcal{G}$ of pairwise nonisomorphic graphs and a point set P that admits a simultaneous embedding of $\mathcal{G}'$. Since $\mathcal{G}'$ is a class of maximal planar graphs, the convex hull of P must be a triangle. For each $G \in \mathcal{G}'$ we can select an outer face $f(G)$ and a mapping $\pi(G)$ for the vertices bounding $f(G)$ to the convex hull of P so that the resulting straight-line embedding, which by Lemma 5 is completely determined by $f(G)$ and $\pi(G)$, is plane.

Let us group the graphs from $\mathcal{G}'$ into bins, according to the maps f and π. For f, there are $7 \cdot 11$ possible choices: one of the eleven faces of one of the seven graphs in $\mathcal{T}$. For π there are three choices: one of the three possible rotations to map the face chosen by f to the convex hull of P. Note that regarding π there is no additional factor of two for the orientation of the face, because by flip-symmetry such a change corresponds to a different graph (for $T_1, \ldots, T_6$) or a different face of the graph (for T_7), that is, a different choice for f. Altogether this yields a partition of $\mathcal{G}'$ into $3 \cdot 77 = 231$ bins.

The crucial observation (and ultimate reason for this subdivision) is that for all graphs in a single bin the vertices of B (the bipyramid) are mapped to the same points. This is a consequence of the uniqueness of the embedding up to the mapping for the outer face (Lemma 5), which is identical for all graphs in the same bin. Therefore, the triangle t of B in which the outer face is located is mapped to the same oriented triple of points in P for all graphs in the same bin. From there the pattern repeats, noting that every face of B contains the same number of points (five) and that the polyhedron B is face-transitive so that there is no difference as to which face of B was selected to contain the outer face.

It follows that for all graphs in the same bin the graphs from $\mathcal{T}$ planted onto the faces of B are mapped to the same point sets. Any two (nonisomorphic) graphs from $\mathcal{G}'$ differ in at least one of those faces – and by definition not in the one in which the outer face was selected by f. In order for the graphs in a bin to be simultaneously embeddable on P, by Lemma 7 there are at most two different graphs from $\mathcal{T}$ mapped to any of the remaining five faces of B. Therefore there cannot be more than $2^5 = 32$ graphs from $\mathcal{G}'$ in any bin. Hence $|\mathcal{G}'| \leq 231 \cdot 32 = 7'392$, as claimed. □

Since there are strictly more nonisomorphic graphs in $\mathcal{G}$ than can possibly be simultaneously embedded, not all graphs of $\mathcal{G}$ admit a simultaneous embedding. In particular, any subset of $7'392 + 1$ nonisomorphic graphs in $\mathcal{G}$ is a collection that does not have a simultaneous embedding. This proves our Theorem 2.

4 Small n-universal Point Sets

As we have seen in the previous sections, there are no n-universal point sets of size n for $n \geq 15$. In this section, we consider the case $n < 15$. Specifically, we used a computer program [16] to show the following:

Theorem 4. *There exist n-universal point sets of size n for all $1 \leq n \leq 10$.*

We use a straightforward brute-force approach. The two main ingredients are the aforementioned order type database [15] with point sets of size $n \leq 10$ and the *plantri* program for generating maximal planar graphs [17,18]. To determine if a point set P of size n is n-universal, our program tests if for all maximal planar graphs $G = (V, E)$ on n vertices, there exists a bijection $\varphi : V \to P$ such that straight-line drawing of G induced by φ is plane. If such a bijection exists for all G, then P is universal. Otherwise, there is a graph G that has no plane straight-line embedding on P. Note that it is sufficient to consider maximal planar graphs since adding edges only makes the embedding problem more difficult. Work on the case $n = 11$ is still in progress at the time of writing. For $n > 11$ the approach unfortunately becomes infeasible; it is unknown whether or not there exist n-universal point sets of size n for $11 \leq n \leq 14$. Table 1 gives an overview of the results of this paper and Fig. 5 shows one universal point set for each $n = 5, \ldots, 10$.

Table 1. The number of (non-equivalent) n-universal point sets of size n

n:	1	2	3	4	5	6	7	8	9	10	11	12	13	14	≥ 15
# universal point sets:	1	1	1	1	1	5	45	364	5′955	2′072	?	?	?	?	0

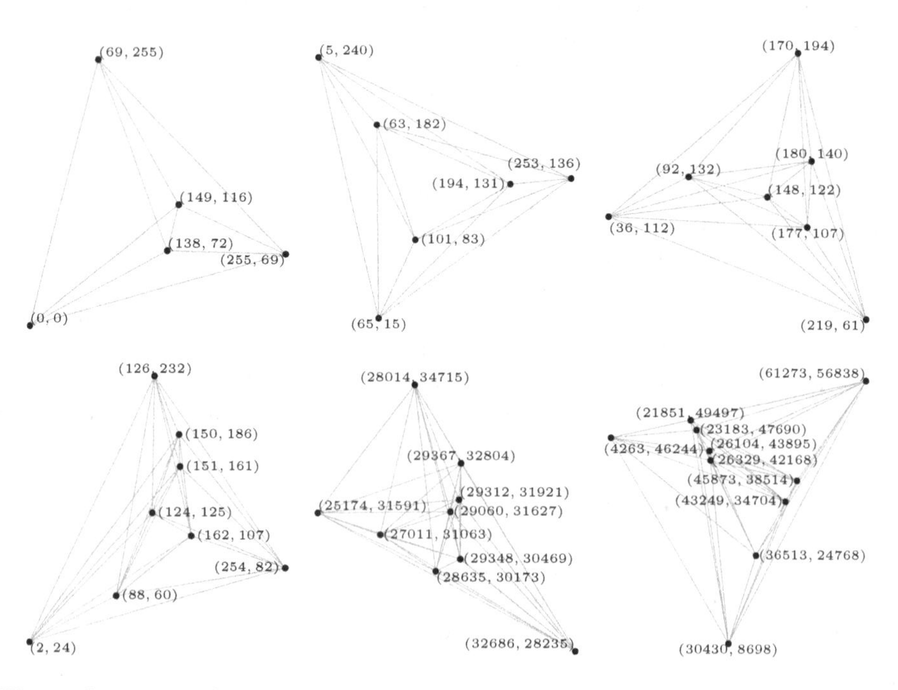

Fig. 5. One universal point set for each $n = 5, \ldots, 10$. Each pair of points is connected with a line segment

References

1. de Fraysseix, H., Pach, J., Pollack, R.: How to draw a planar graph on a grid. Combinatorica 10(1), 41–51 (1990)
2. Schnyder, W.: Embedding planar graphs on the grid. In: Proc. 1st ACM-SIAM Sympos. Discrete Algorithms, pp. 138–148 (1990)
3. Kurowski, M.: A 1.235 lower bound on the number of points needed to draw all n-vertex planar graphs. Information Processing Letters 92(2), 95–98 (2004)
4. Chrobak, M., Karloff, H.J.: A lower bound on the size of universal sets for planar graphs. SIGACT News 20(4), 83–86 (1989)
5. Demaine, E.D., Mitchell, J.S.B., O'Rourke, J.: The Open Problems Project, Problem #45, `http://maven.smith.edu/~orourke/TOPP/P45.html`
6. Kobourov, S.G.: Personal communication (2012)
7. Ábrego, B.M., Fernández-Merchant, S.: A lower bound for the rectilinear crossing number. Graphs and Combinatorics 21(3), 293–300 (2005)
8. Lovász, L., Vesztergombi, K., Wagner, U., Welzl, E.: Convex quadrilaterals and k-sets. In: Pach, J. (ed.) Towards a Theory of Geometric Graphs. Contemporary Mathematics, vol. 324, pp. 139–148. American Mathematical Society, Providence (2004)
9. Fulek, R., Tóth, C.D.: Universal point sets for planar three-trees. CoRR abs/1212.6148 (2012)
10. Fáry, I.: On straight lines representation of planar graphs. Acta Sci. Math. Szeged 11, 229–233 (1948)
11. Wagner, K.: Bemerkungen zum Vierfarbenproblem. Jahresbericht der Deutschen Mathematiker-Vereinigung 46, 26–32 (1936)
12. Beineke, L.W., Pippert, R.E.: Enumerating dissectible polyhedra by their automorphism groups. Canad. J. Math. 26, 50–67 (1974)
13. Brass, P., Cenek, E., Duncan, C.A., Efrat, A., Erten, C., Ismailescu, D.P., Kobourov, S.G., Lubiw, A., Mitchell, J.S.: On simultaneous planar graph embeddings. Comput. Geom. Theory Appl. 36(2), 117–130 (2007)
14. Goodman, J.E., Pollack, R.: Multidimensional sorting. SIAM J. Comput. 12(3), 484–507 (1983)
15. Aichholzer, O., Krasser, H.: The point set order type data base: A collection of applications and results. In: Proc. 13th Canad. Conf. Comput. Geom., Waterloo, Canada, pp. 17–20 (2001)
16. Cardinal, J., Hoffmann, M., Kusters, V.: A program to find all universal point sets (2013), `http://people.inf.ethz.ch/kustersv/universal.html`
17. Brinkmann, G., McKay, B.: Fast generation of planar graphs. MATCH Communications in Mathematical and in Computer Chemistry 58(2), 323–357 (2007)
18. Brinkmann, G., McKay, B.: The program plantri (2007), `http://cs.anu.edu.au/~bdm/plantri`

On Non 3-Choosable Bipartite Graphs

W. Charoenpanitseri[1], N. Punnim[2], and C. Uiyyasathian[1,*]

[1] Department of Mathematics and Computer Science, Faculty of Science, Chulalongkorn University, Bangkok, 10330, Thailand
[2] Department of Mathematics, Srinakharinwirot University, Sukhumvit 23, Bangkok 10110, Thailand
ch_wongsakorn@hotmail.com, punnim@gmail.com, chariya.u@chula.ac.th

Abstract. In 2003, Fitzpatrick and MacGillivray proved that every complete bipartite graph with fourteen vertices except $K_{7,7}$ is 3-choosable and there is the unique 3-list assignment L up to renaming the colors such that $K_{7,7}$ is not L-colorable. We present our strategies which can be applied to obtain another proof of their result. These strategies are invented to claim a stronger result that every complete bipartite graph with fifteen vertices except $K_{7,8}$ is 3-choosable. We also show all 3-list assignments L such that $K_{7,8}$ is not L-colorable.

Keywords and Phrases: list coloring, choosability.

1 Introduction

A *list assignment* of a graph G is a mapping which assigns a set of colors, called a *list*, to each vertex of G. A *k-list assignment* of G is a list assignment L such that $|L(v)| = k$ for every $v \in V(G)$. A *coloring* of G is a mapping from $V(G)$ to a set of colors such that endpoints of each edge has different colors. Given a list assignment L, a coloring f of G is an *L-coloring* of G if $f(v)$ is chosen from $L(v)$ for every vertex v of G. If a graph G has an L-coloring, we say that G is *L-colorable* or L is a *colorable list assignment* of G. A graph G is *k-choosable* if G is L-colorable for every k-list assignment L. Let $S \subseteq V(G)$. If L is a list assignment of G, we let $L|_S$ denote L restricted to S and $L(S)$ denote $\bigcup_{v \in S} L(v)$. Let $\mathcal{F} = \{\{1,2,3\},\{1,4,5\},\{1,6,7\},\{2,4,6\},\{2,5,7\},\{3,4,7\},\{3,5,6\}\}$. It is well-known that $\mathcal{F}$ is the collection of lines of the Fano plane which is unique up to renaming the colors. $L_{\mathcal{F}}$ denotes the 3-list assignment of $K_{7,7}$ such that all seven vertices in each partite set are assigned by distinct lists from $\mathcal{F}$.

The problem of list assignments was first studied by Vizing [8] and by Erdős, Rubin and Taylor [2]. In both papers, the authors proved that there exists a non k-choosable bipartite graph for every positive number k. For example, $K_{7,7}$ is not 3-choosable, and $L_{\mathcal{F}}$ is a non 3-colorable list assignment of $K_{7,7}$. Moreover, in [2], the authors give a characterization of 2-choosable graphs. However, for $k \geq 3$, there is no literature giving a characterization of k-choosable graphs in general, only some specific classes of graphs are investigated. For example, every planar

* Corresponding author.

J. Akiyama, M. Kano, and T. Sakai (Eds.): TJJCCGG 2012, LNCS 8296, pp. 42–56, 2013.

graph is 5-choosable, while some authors found some classes of 3-choosable planar graphs.(See [5],[6],[7],[10],[11],[12],[13].) Recently, (k,t)-choosability of graphs is explored in [1]. In 1996, Hanson, MacGillivray and Toft[4] stated that every complete bipartite graph with thirteen vertices is 3-choosable. In other word, $K_{7,7}$ is the smallest non 3-choosable graph. Later, Fitzpatrick and MacGillivray [3] added that every complete bipartite graph with fourteen vertices except $K_{7,7}$ is 3-choosable. Furthermore, $L_{\mathcal{F}}$ is the unique 3-list assignment up to renaming the colors which is a non 3-colorable list assignment of $K_{7,7}$. In this paper, we explore their results to show that every complete bipartite graph with fifteen vertices is 3-choosable except $K_{7,8}$, and a 3-list assignment L of $K_{7,8}$ is a non-colorable list assignment if and only if $L|_{V(K_{7,7})} = L_{\mathcal{F}}$. However, the proof in [3] does not seem to be extendable, we then construct new strategies. These strategies not only provide our main result, but also give another proof of [3].

Let L be a list assignment of the complete bipartite graph $K_{a,b}$. The notation L_a and L_b denote the collections of lists assigned to the vertices in the partite sets with a and b vertices, respectively. If $a = b$, we use the notation $L_{a(i)}$ and $L_{a(ii)}$. Given a collection of lists $\mathbb{X} = \{X_1, X_2, \ldots, X_n\}$, a *coloring* of $\mathbb{X}$ is a set $C \subseteq X_1 \cup X_2 \cup \ldots \cup X_n$ such that $C \cap X_i \neq \emptyset$ for all $i = 1, 2, \ldots, n$. A coloring C of X is called a *t-coloring* if $|C| = t$.

Lemma 1. [3],[4] *Let L be a list assignment of the complete bipartite graph $K_{a,b}$. Then $K_{a,b}$ is not L-colorable if and only if every coloring of L_a(or L_b) has a subset that is a list in L_b(or L_a, respectively).*

Theorem 1. [3],[4] *Let G be an n-vertex graph. If G is L_1-colorable for every k-list assignment L_1 such that $|\bigcup_{v \in V(G)} L_1(v)| = t$ and $n\binom{k}{2} < \binom{t+1}{2}$, then G is L_2-colorable for every k-list assignment L_2 such that $|\bigcup_{v \in V(G)} L_2(v)| \geq t$.*

2 Strategies

To prove the main result, many similar cases are considered. Thus we construct tools to deal with each case. The first tool is for the cases that all lists assigned to the vertices in one partite set are mutually disjoint.

Strategy A. *Let L be a list assignment of $K_{a,b}$ with $L_a = \{A_1, A_2, \ldots, A_a\}$, $L_b = \{B_1, B_2, \ldots, B_b\}$ and all lists have size at most 3. If all lists in L_a are mutually disjoint and $\prod_{i=1}^{a} |A_i| > 3^{a-1}n_1 + \lfloor 3^{a-2} \rfloor n_2 + \lfloor 3^{a-3} \rfloor n_3$ where $n_i = |\{B \in L_b, |B| = i\}|$ for $i = 1, 2, 3$, then $K_{a,b}$ is L-colorable.*

Proof. Since there are $|A_i|$ possible ways to color the list A_i for each i and all A_i's are mutually disjoint, the number of a-colorings of L_a is $\prod_{i=1}^{a} |A_i|$. We count the number of those a-colorings containing each B_i of L_b for $i = 1, 2, \ldots, b$. Consider $B_i \in L_b$.

Case 1. $|B_i| = 1$, *say* $B_i = \{r\}$.
If $r \notin A_j$ for all $j = 1, 2, \ldots, a$, then all a-colorings of L_a do not contain B_i. Without loss of generality, suppose that $r \in A_1$. To complete an a-coloring of L_a,

we choose the other $a-1$ colors each from the remaining A_j where $j = 2, 3, \ldots, a$. Thus the number of the a-colorings of L_a containing r is $\prod_{j=2}^{a} |A_j|$. That is, the number of the a-colorings of L_a which contain B_i as a subset is at most 3^{a-1}.

Case 2. $|B_i| = 2$, *say* $B_i = \{r, s\}$.
Consider an a-coloring of L_a containing both r and s. Without loss of generality, suppose that $r \in A_1$ and $s \in A_2$. To complete an a-coloring of L_a, we choose the other $a-2$ colors each from the remaining A_j where $j = 3, 4, \ldots, a$. Thus the number of the a-colorings of L_a which contain B_i as a subset is $\prod_{j=3}^{a} |A_j|$. That is, the number of the a-colorings of L_a which contain B_i as a subset is at most 3^{a-2}. Note that in case $a = 1$, all a-colorings are 1-colorings; hence, the number of a-colorings containing B_i as a subset is $\lfloor 3^{a-2} \rfloor = 0$.

Case 3. $|B_i| = 3$, *say* $B_i = \{r, s, t\}$.
Consider an a-coloring of L_a containing r, s and t. Without loss of generality, suppose that $r \in A_1, s \in A_2, t \in A_3$. Again, we can choose the other $a-3$ colors from each A_j where $j = 4, 5, \ldots, a$. Thus the number of the a-colorings of L_a which contain B_i as a subset is $\prod_{j=4}^{a} |A_j|$. That is, the number of the a-colorings of L_a which contain B_i as a subset is at most 3^{a-3}. Note that in the case $a \leq 2$, all a-colorings are 1-colorings or 2-colorings; hence, the number of a-colorings containing B_i as a subset is $\lfloor 3^{a-3} \rfloor = 0$.

Hence L_a has at most $3^{a-1}n_1 + \lfloor 3^{a-2} \rfloor n_2 + \lfloor 3^{a-3} \rfloor n_3$ a-colorings containing some B_i. Since the number of a-colorings of L_a is $\prod_{j=1}^{a} |A_j|$ and $\prod_{j=1}^{a} |A_j| > 3^{a-1}n_1 + \lfloor 3^{a-2} \rfloor n_2 + \lfloor 3^{a-3} \rfloor n_3$, there exists a coloring of L_a which does not contain any list in L_b. Therefore, $K_{a,b}$ is L-colorable.

The same result can be concluded if we consider the other way around, that is, the assumption in Strategy A for a list assignment L of $K_{a,b}$ becomes the statement that *all lists in* L_b *are mutually disjoint and* $\prod_{i=1}^{a} |B_i| > 3^{b-1}n_1 + \lfloor 3^{b-2} \rfloor n_2 + \lfloor 3^{b-3} \rfloor n_3$, *where* $n_i = |\{A \in L_a, |A| = i\}|$ *for* $i = 1, 2, 3$. In this case, we call *Strategy A for* L_b and we call the original version *Strategy A* for L_a.

Note that the value $3^{a-1}n_1 + 3^{a-2}n_2 + 3^{a-3}n_3$ is sharp because of a non-colorable 3-list assignment L of $K_{3,27}$ such that $L_3 = \{\{1, 2, 3\}, \{4, 5, 6\}, \{7, 8, 9\}\}$ and $L_{27} = \{\{a, b, c\} | a \in \{1, 2, 3\}, b \in \{4, 5, 6\}, c \in \{7, 8, 9\}\}$.

The next five strategies are tools to find an L-coloring of $K_{a,b}$ with respect to a list assignment L in the case that a color appears in at least $a-1, a-2, a-3, a-4$ and $a-5$ lists in L_a, respectively. The next strategy is called *Strategy B for* L_a and we can define *Strategy B* for L_b, similarly.

Strategy B. *Let* L *be a 3-list assignment of* $K_{a,b}$. *If a color appears in* $a-1$ *lists in* L_a, *then* $K_{a,b}$ *is* L*-colorable.*

Proof. Because a color appears in $a-1$ lists in L_a, we can label L_a by at most two colors. Since every list in L_b has size 3, all lists in L_b still have available colors.

Remark 1. Let L be a list assignment of $K_{a,b}$ such that L_a has a 2-coloring C. Then,

(i) if L is a 3-list assignment then $K_{a,b}$ is L-colorable;

(ii) if all lists of size at most 2 in L_b have a color which is not in C, then $K_{a,b}$ is L-colorable.

Strategy C. *Let L be a 3-list assignment of $K_{a,b}$ such that every color appears in at most eight lists in L_b. If a color appears in $a-2$ lists in L_a then $K_{a,b}$ is L-colorable.*

Proof. Strategy B takes care the case that a color appears in more than $a-2$ lists in L_a. Assume that a color appears in exactly $a-2$ lists in L_a. If the two remaining lists in L_a have a common color, then there exists a 2-coloring of L_a. Since all lists in L_b are of size 3, $K_{a,b}$ is L-colorable by Remark 1. Suppose that the two remaining lists in L_a have no common color. Hence, L_a has at least nine 3-colorings containing color 1. However, by the assumption, color 1 appears in at most eight lists in L_b. Thus, at least one of such nine 3-colorings is not a list in L_b. Therefore, by Lemma 1, $K_{a,b}$ is L-colorable.

For convenience, from now on, we write lists without commas and braces. For example, $\{123, 145, 167, 246, 257, 347, 356\}$ is written instead of $\{\{1,2,3\}, \{1,4,5\}, \{1,6,7\}, \{2,4,6\}, \{2,5,7\}, \{3,4,7\}, \{3,5,6\}\}$. For a list A, the notation $A-1$ represents the list which is obtained from A by removing color 1 from A. Similarly, the notation $A-12$ represents the list which is obtained from A by removing color 1 and color 2 from A.

Strategy D. *Let L be a 3-list assignment of $K_{a,b}$ such that every color appears in at most r lists in L_b. If there exists a color that appears in at least $a-3$ lists in L_a and $(r,b) \in \{(r,b)|r \leq 2, b \leq 22\} \cup \{(3,b)|b \leq 14\} \cup \{(4,b)|b \leq 12\} \cup \{(5,b)|b \leq 9\}$, then $K_{a,b}$ is L-colorable.*

Proof. Let $L_a = \{A_1, A_2, \ldots, A_a\}$ and $L_b = \{B_1, B_2, \ldots, B_b\}$. If there is a color appearing in more than $a-3$ lists, we apply Strategy C. Assume that color 1 appears in exactly $a-3$ lists in L_a, say $1 \in A_1, A_2, \ldots, A_{a-3}$. First, we label $A_1, A_2, \ldots, A_{a-3}$ by color 1. Now, we consider the remaining vertices which form $K_{3,b}$. For the worst case, we may suppose that $1 \in B_1, B_2, \ldots, B_r$. Let L' be the list assignment of $K_{3,b}$ which is obtained from L by removing color 1. Notice that $L'_3 = \{A_{a-2}, A_{a-1}, A_a\}$ and $L'_b = \{B_1 - 1, \ldots, B_r - 1, B_{r+1} \ldots, B_b\}$. If $A_{a-2} \cap A_{a-1} \cap A_a \neq \emptyset$ then there is a 2-coloring of L_a ; hence, $K_{a,b}$ is L-colorable by Remark 1. Suppose that $A_{a-2} \cap A_{a-1} \cap A_a = \emptyset$.

Case 1. $|A_{a-2} \cap A_{a-1}| = 2$.
Let $2, 3 \in A_{a-2}, A_{a-1}$ and $A_a = 456$. Then L_a has at least six 3-colorings, called $\{1,2,4\}, \{1,2,5\}, \{1,2,6\}, \{1,3,4\}, \{1,3,5\}, \{1,3,6\}$. Since $r \leq 5$, at least one of the six 3-colorings is not a list in L_b. By Lemma 1, $K_{a,b}$ is L-colorable.

Case 2. $|A_{a-2} \cap A_{a-1}| = 1$.
Let $A_{a-2} = 234, A_{a-1} = 256$ and $A_a = pqr$ where $p, q, r \notin \{1,2\}$. There are several subcases.

Case 2.1 $\{p,q,r\} \cap \{3,4,5,6\} \neq \emptyset$.
Without loss of generality, we let $p = 3$. Then L_a has at least five 3-colorings, called $\{1,2,3\}, \{1,2,q\}, \{1,2,r\}, \{1,3,5\}, \{1,3,6\}$. If one of such 3-colorings is

not a list in L_b, then $K_{a,b}$ is L-colorable by Lemma 1. Suppose that such 3-colorings are lists in L_b. Thus $r = 5$ and $b \leq 9$. Let $B_1 = 123, B_2 = 12q, B_3 = 12r, B_4 = 135$ and $B_5 = 136$. We label B_1, B_2, B_3, B_4 and B_5 by color 2 and color 3. Now, the remaining vertices form a $K_{3,b-5}$ where $b \leq 9$. For the worst case, we suppose $b = 9$. Let L'' be the list assignment of $K_{3,4}$ which is obtained from L' by removing color 2. Then $L''_3 = \{4, 56, qr\}$ and $L''_4 = \{B_6, B_7, \ldots, B_9\}$. If L''_3 has a 2-coloring, then $K_{3,4}$ is L''-colorable by Remark 1. Hence, suppose that L''_3 has no 2-coloring. That is, $q, r \notin \{4, 5, 6\}$. We let $q = 7$ and $r = 8$. Then L''_3 has four 3-colorings, namely $\{4,5,7\}, \{4,5,8\}, \{4,6,7\}, \{4,6,8\}$. Again, we suppose that such 3-colorings are lists in L''_4. Now, $L_b = L_9 = \{123, 127, 128, 135, 136, 457, 458, 478, 468\}$. Hence, color 1 and color 4 form a 2-coloring of L_b. By Remark 1, $K_{a,b}$ is L-colorable.

Case 2.2 $p, q, r \notin \{3, 4, 5, 6\}$.

Let $p = 7, q = 8$ and $r = 9$. Then $\{1,2,7\}, \{1,2,8\}$ and $\{1,2,9\}$ are 3-colorings of L_a. Again, by Lemma 1, $K_{a,b}$ is L-colorable unless the case that L_b contains 127, 128 and 129. Let $B_1 = 127, B_2 = 128, B_3 = 129$. Thus $r \geq 3$. Next, we label B_1, B_2, B_3 by color 2. Let L'' be the list assignment of $K_{3,b-3}$ which is obtained from L' by removing color 2. Then $L''_3 = \{A_{a-2} - 2, A_{a-1} - 2, A_a\}$ and $L''_{b-3} = \{B_4 - 1, \ldots, B_r - 1, B_{r+1}, B_{r+2}, \ldots, B_b\}$. Now, we apply Strategy A for L''_3.

Case 2.2.1 $r = 3$.

Then all lists in L''_{b-3} are of size 3. Apply Strategy A for L''_3 because $12 > 3^{3-3}(b-3)$.

Case 2.2.2 $r = 4$.

For the worst case, we suppose that $1 \in B_4$. That is, L''_{b-3} has exactly one lists of size 2 and the remaining lists are of size 3. Again, we apply Strategy A for L''_3 because $12 > 3^{3-2} \cdot 1 + 3^{3-3}(b-4)$.

Case 2.2.3 $r = 5$.

For the worst case, we suppose that $1 \in B_4, B_5$. That is, L''_{b-3} has exactly two lists of size 2 and the remaining lists are of size 3. Again, we apply Strategy A for L''_3 because $12 > 3^{3-2} \cdot 2 + 3^{3-3}(b-5)$.

Case 3. *A_{a-2}, A_{a-1}, A_a are mutually disjoint.*

Then $|A_{a-2}| \cdot |A_{a-1}| \cdot |A| = 3^3$ Now, we use Strategy A for L'_3. Note that there are r lists in L_b containing color 1. So the number of lists of size 2 and size 3 in L'_b are $n_2 = r$ and $n_3 = b - r$, respectively. Thus $3 \cdot r + (b - r) < 3^3$. Hence $K_{3,b}$ is L'-colorable by Strategy A for L'_3. Therefore, $K_{a,b}$ is L-colorable.

Lemma 2. *Let L be a 3-list assignment of $K_{a,b}$ such that every color appears in at most three lists in L_b. If color 1 appears in exactly $a - 4$ lists in L_a and colors 1 and 2 appear together in three lists in L_b, then $K_{a,b}$ is L-colorable.*

Proof. Let $L_a = \{A_1, A_2, \ldots, A_a\}$ and $L_b = \{B_1, B_2, \ldots, B_b\}$. Assume that $1 \in A_1, A_2, \ldots, A_{a-4}$ and $1, 2 \in B_1, B_2, B_3$. If $A_{a-3} \cap A_{a-2} \cap A_{a-1} \cap A_a$ is not empty, then L_a has a 2-coloring; hence, $K_{a,b}$ is L-colorable by Remark 1. Suppose that $A_{a-3} \cap A_{a-2} \cap A_{a-1} \cap A_a = \emptyset$. Then we label $A_1, A_2, \ldots, A_{a-4}$ by color 1 and label B_1, B_2, B_3 by color 2. Next, consider the remaining vertices which form $K_{4,b-3}$. Let L' be the list assignment of $K_{4,b-3}$ which is obtain from L by removing color 1 and color 2. For the worst case, we suppose that $2 \in A_{a-3}, A_{a-2}, A_{a-1}$. That is, $L'_4 = \{A_{a-3} - 2, A_{a-2} - 2, A_{a-1} - 2, A_a\}$ and $L'_{b-3} = \{B_4, B_5, \ldots, B_b\}$. If any two lists in L'_4 have a common color, it can be verified that L'_4 has at least four 3-colorings of L'_4. Since every color appears in at most three lists in L'_{b-3}, at least one of these 3-colorings is not a list in L'_{b-3}. Then we suppose that all lists in L'_4 have no common color. Let $L'_4 = \{34, 56, 78, 9AB\}$. Since all lists in L'_4 are subsets of $\{3, 4, 5, 6, 7, 8, 9, A, B\}$, we may suppose that all lists in L'_{b-3} are subsets of $\{3, 4, 5, 6, 7, 8, 9, A, B\}$. Since every color appears in at most three lists in L'_b, we obtain $b - 3 \leq 9$.

Case 1. $b - 3 \leq 7$.
Then $K_{4,b-3}$ is L'-colorable by Strategy A for L'_4.

Case 2. $b - 3 = 8$.
We consider the possibility of L'_8 such that $K_{4,8}$ is not L'-colorable. Then L'_8 must be $\{357, 358, 367, 368, 457, 458, 467, 468\}$. However, this case cannot occur because every color appears in at most three lists in L'_8.

Case 3. $b - 3 = 9$.
Then every color from $3, 4, 5, 6, 7, 8, 9, A, B$ must appear in three lists in L'_9. We label 34 in L'_4 by color 3 and label three lists containing color 4 in L'_9 by color 4. The remaining vertices form a $K_{3,6}$. Let L'' be the list assignment of $K_{3,6}$ which is obtained from L' by removing color 3 and color 4. Then $L''_3 = \{56, 78, 9AB\}$. For the worst case, we suppose that L''_6 has three lists of size 2 and three lists of size 3. Again, we consider the possibilities of L''_6 such that $K_{4,6}$ is not L''-colorable. Without loss of generality, L''_6 must be $\{57, 58, 67, 689, 68A, 68B\}$. However, this case cannot occur because every color appears in at most three lists in L''_6.

Strategy E. *Let L be a 3-list assignment of $K_{a,b}$ such that every color appears in at most r lists in L_b. If color 1 appears in $a - 4$ lists in L_a, and $(r, b) \in \{(r, b) | r \leq 2, b \leq 22\} \cup \{(3, b) | b \leq 14\}$, then $K_{a,b}$ is L-colorable unless the four remaining lists in L_a are* 246, 257, 347, 356 *and $\mathcal{F} \subseteq L_b$.*

Proof. Let $L_a = \{A_1, A_2, \ldots, A_a\}$ and $L_b = \{B_1, B_2, \ldots, B_b\}$. If there is a color appearing in more than $a-4$ lists in L_a, then we apply Strategy D. Assume that color 1 appears in exactly $a-4$ lists in L_a, say $1 \in A_1, A_2, \ldots, A_{a-4}$. Moreover, we suppose that the four remaining lists in L_a are not 246, 257, 347, 356 or $\mathcal{F} \not\subseteq L_b$.

We first label $A_1, A_2, \ldots, A_{a-4}$ by color 1. Then the remaining vertices form $K_{4,b}$. For the worst case, we may suppose that $1 \in B_1, B_2, \ldots, B_r$. Let L' be the list assignment of $K_{4,b}$ which is obtained from L by removing color 1. Then $L'_4 = \{A_{a-3}, A_{a-2}, A_{a-1}, A_a\}$ and $L'_b = \{B_1 - 1, \ldots, B_r - 1, B_{r+1}, \ldots, B_b\}$.

Case 1. *There is a color appearing in all lists in L'_4.*
Thus we use such color to label all lists in L'_4. It follows that every list in L'_b still has an available color. Then $K_{a,b}$ is L-colorable.

Case 2. *There is a color appearing in three lists in L'_4.*
If a color appears in four lists, then it is done by case 1. Suppose that no color appears in four lists in L'_4. Let $2 \in A_{a-3} \cap A_{a-2} \cap A_{a-1}$ and $A_a = 345$. Now, we consider L of $K_{a,b}$. Then L_a has at least three 3-colorings, that is, $\{1, 2, 3\}, \{1, 2, 4\}, \{1, 2, 5\}$. If L_b does not contain all of these 3-colorings, $K_{a,b}$

is immediately L-colorable by Lemma 1. Otherwise, we suppose that $B_1 = 123, B_2 = 124, B_3 = 125$. By Lemma 2, $K_{a,b}$ is L-colorable.

Case 3. *There is a color appearing in two lists in L'_4 and the remaining two lists have no common color.*

If there is a color appearing in more than two lists, then the proof is done by Case 1 and Case 2. Suppose that each color appears in at most two lists in L'_4. Let $2 \in A_{a-3}, A_{a-2}$ and $A_{a-1} \cap A_a = \emptyset$. We next label A_{a-3} and A_{a-2} by color 2. Then, we focus on the remaining vertices which form a $K_{2,b}$. Let L'' be the list assignment of $K_{2,b}$ which is obtained from L' by removing color 2. Since we use color 1 and color 2 to label lists in L_a, we may suppose that both color 1 and color 2 appear in three lists in L''_b for the worse case. Thus, there are four possibilities of L''_b.

Case 3.1 *L''_b has six lists of size* 2 *and* $b-6$ *lists of size* 3.

We see that $|A_{a-1}|\cdot|A_a| = 3^2 > 3^0\cdot 6$. By Strategy A for L''_2, $K_{2,b}$ is L''-colorable. Then $K_{a,b}$ is L-colorable.

Case 3.2 *L''_b has one list of size* 1, *four lists of size* 2 *and* $b-5$ *lists of size* 3.

We see that $|A_{a-1}| \cdot |A_a| = 3^2 > 3 \cdot 1 + 4$. By Strategy A for L''_2, $K_{2,b}$ is L''-colorable. Then $K_{a,b}$ is L-colorable.

Case 3.3 *L''_b has two lists of size* 1, *two lists of size* 2 *and* $b-4$ *lists of size* 3.

We see that $|A_{a-1}| \cdot |A_a| = 3^2 > 3 \cdot 2 + 2$. By Strategy A, $K_{2,b}$ is L''-colorable. Then $K_{a,b}$ is L-colorable.

Case 3.4 *L''_b has three lists of size* 1, *no list of size* 2 *and* $b-3$ *lists of size* 3.

That is, color 1 and color 2 appear together in exactly three lists of L_b. Then $K_{a,b}$ is L-colorable by Lemma 2.

Case 4. *There is a color appearing in two lists in L'_4 and the remaining two lists have a common color.*

Similar to case 3, we suppose that no color appears in three lists in L'_4. Let $2 \in A_{a-3}, A_{a-2}$ and $3 \in A_{a-1} \cap A_a$. Hence, $\{1,2,3\}$ is a 3-coloring of L_a. If 123 is not a list in L_b, then $K_{a,b}$ is L-colorable by Lemma 1. Otherwise, we suppose that $B_1 = 123$.

Case 4.1 $|A_{a-3} \cap A_{a-2}| \geq 2$ *and* $|A_{a-1} \cap A_a| \geq 2$.

Let $4 \in A_{a-3} \cap A_{a-2}$ and $5 \in A_{a-1} \cap A_a$. We obtain at least four 3-colorings of L_a, that is, $\{1,2,3\}, \{1,2,5\}, \{1,4,3\}, \{1,4,5\}$. Since each color appears in at most three lists in L_b, at least one of such 3-colorings is not a list in L_b. Then $K_{a,b}$ is L-colorable by Lemma 1.

Case 4.2 $|A_{a-3} \cap A_{a-2}| \geq 2$ *and* $|A_{a-1} \cap A_a| = 1$.

We may suppose that $|A_{a-3} \cap A_{a-2}| = 2$. Let $A_{a-3} = 24x, A_{a-2} = 24y, A_{a-1} = 356$ and $A_a = 378$ where $x \neq y$ and $x, y \notin \{1,2,3,4\}$. Then $\{1,4,3\}$ is a 3-coloring of L_a. If 143 is not a list in L_b, then $K_{a,b}$ is L-colorable by Lemma 1. Otherwise, suppose that $B_2 = 143$. Recall that we have already labeled $A_1, A_2, \ldots, A_{a-4}$ by color 1. Now, we label B_1, B_2 by color 3. Consider the uncolor vertices which form a $K_{4,b-2}$. Let L'' be a list assignment of $K_{4,b-2}$ which is obtained from L by removing color 3. Then $L''_4 = \{24x, 24y, 56, 78\}$ and $L''_{b-2} = \{B_3 - 1, B_4, B_5, \ldots, B_b\}$. By the fact that L'_4 has at least eight 3-colorings and every color appears in at most three colors in L_b, it can be verified that $K_{4,b-2}$ is L''-colorable.

Case 4.3 $|A_{a-3} \cap A_{a-2}| = 1$ *and* $|A_{a-1} \cap A_a| \geq 2$.
It is similar to Case 4.2.

Case 4.4 $|A_{a-3} \cap A_{a-2}| = 1$ *and* $|A_{a-1} \cap A_a| = 1$.
Let $A_{a-1} = 345, A_a = 367$ and $A_{a-3} = 2ef, A_{a-2} = 2gh$ where e, f, g, h are distinct. Note that $\{1,2,4,6\}, \{1,2,4,7\}, \{1,2,5,6\}, \{1,2,5,7\}$ are 4-colorings of L_a. By Lemma 1, if one of these 4-colorings has no subset that is a list in L_b, then $K_{a,b}$ is L-colorable. Again, suppose that these 4-colorings have a subset that is a list in L_b. Without loss of generality, L_b can be verified that there are two possibilities of L_b.

Case 4.4.1 $B_2 = 124$ *and* $B_3 = 125$.
Then $K_{a,b}$ is L-colorable by Lemma 2.

Case 4.4.2 $B_2 = 146, B_3 = 147, B_4 = 256, B_5 = 257$.
Recall that we have already labeled $A_1, A_2, \ldots, A_{a-4}$ by color 1. Now, we label A_{a-1}, A_a by color 3 and label B_1, B_4, B_5 by color 2. Next, consider the remaining vertices which forms a $K_{2,b-3}$. Let L'' be the list assignment of $K_{2,b-3}$ which is obtained from L' by removing color 2 and color 3. That is, $L''_2 = \{ef, gh\}$ and $L''_{b-3} = \{46, 47, B_6, B_7, \ldots, B_b\}$. Then L''_2 has exactly four 2-colorings, namely $\{e,g\}, \{e,h\}, \{f,g\}$ and $\{f,h\}$. If one of such 2-colorings is not a list in L''_{b-3}, then $K_{2,b-3}$ is L''-colorable by Lemma 1. Suppose that such four 2-colorings are lists in L''_{b-3}. Then L''_{b-3} has at least four lists of size 2. Recall that $3 \in B_1$. Then color 3 appears in two lists in $B_6, B_7, \ldots, B_b$. Hence, we suppose that $3 \in B_6, B_7$. Then $L''_{b-3} = \{56, 57, B_6 - 3, B_7 - 3, B_8, B_9, \ldots, B_{b-3}\}$.

Let L^* be a 2-list assignment of $K_{2,4}$ such that $L^*_2 = \{ef, gh\}$ and $L^*_4 = \{56, 57, B_6 - 3, B_7 - 3\}$. By Remark 1, $K_{2,b-3}$ is L''-colorable if and only if $K_{2,4}$ is L^*-colorable. Moreover, $K_{2,4}$ is not L^*-colorable if and only if $L^*_2 = \{45, 67\}$ and $L^*_4 = \{46, 47, 56, 57\}$. Therefore, $K_{2,4}$ is not L^*-colorable if and only if $\{A_{a-3}, A_{a-2}, A_{a-1}, A_a\} \neq \{246, 257, 347, 356\}$ or $\mathcal{F} \not\subseteq L_b$.

Case 5. *All lists in* L'_4 *are mutually disjoint.*
Note that L'_b has $b - r$ lists of size 3, r lists of size 2 and no list of size 1. We have that $\prod_{i=a-3}^{a} |A_i| = 3^4 > 3^2 \cdot r + 3 \cdot (b - r)$. By Strategy A for L'_4, $K_{4,b}$ is L'-colorable.

Lemma 3. *Let L be a 3-list assignment of $K_{a,b}$ such that every color appears in at most two lists in L_b. If a color appears in exactly $a - 5$ lists in L_a and a color appears in exactly three of the five remaining lists, then $K_{a,b}$ is L-colorable.*

Proof. Let $L_a = \{A_1, A_2, \ldots, A_a\}$ and $L_b = \{B_1, B_2, \ldots, B_b\}$. Suppose that $1 \in A_1, A_2, \ldots, A_{a-5}$ and $2 \in A_{a-4}, A_{a-3}, A_{a-2}$. Then we first label $A_1, A_2, \ldots, A_{a-5}$ by color 1 and label $A_{a-4}, A_{a-3}, A_{a-2}$ by color 2. Consider the remaining vertices which form $K_{2,b}$. Let L' be the list assignment of $K_{2,b}$ which is obtained from L by removing color 1 and color 2. Note that $L'_2 = \{A_{a-1}, A_a\}$. The proof is divided into four cases.

Case 1. $A_{a-1} \cap A_a = \emptyset$.
To apply Strategy A for L'_2, we count the number of the lists of size 1, size 2

and size 3 in L'_b. There are three possibilities. Denote that n_i is the number of the lists of size i in L'_b for $i = 1, 2, 3$.

1. $n_1 = 2, n_2 = 0$ and $n_3 = b - 2$.
2. $n_1 = 1, n_2 = 2$ and $n_3 = b - 3$.
3. $n_1 = 0, n_2 = 4$ and $n_3 = b - 4$.

All possibilities satisfy the condition in Strategies A of L'_2. Therefore, $K_{2,b}$ is L'-colorable.

Case 2. $|A_{a-1} \cap A_a| = 1$.
Let $A_{a-1} = 345$ and $A_a = 367$. If 123 is not a list in L_b, then $K_{a,b}$ is L-colorable. Without loss of generality, suppose that $B_1 = 123$. Then we label all lists containing color 3 in L'_b by color 3. Now, we consider all uncolored vertices. For the worst case, we suppose that no other list except B_1 contains color 3. Thus the remaining vertices form $K_{2,b-1}$. Let L'' be the list assignment of $K_{2,b-1}$ which is obtained from L' by removing color 3. Then we can apply Strategy A for L'_2.

Case 3. $|A_{a-1} \cap A_a| = \emptyset$.
Let $A_{a-1} = 345$ and $A_a = 346$. If 123 and 124 are not lists in L_b, then $K_{a,b}$ is immediately L-colorable. Without loss of generality, suppose that $B_1 = 123$ and $B_2 = 124$. Then we label B_1, B_2, A_{a-1} and A_a by color 3, color 4, color 5 and color 6, respectively. Notice that every uncolored vertex in L'_b still has an available color. Therefore, $K_{2,b}$ is L'-colorable.

Strategy F. *Let L be a 3-list assignment of $K_{a,b}$ such that every color appears in at most two lists in L_b. If a color appears in $a - 5$ lists in L_a and $a + b \leq 18$, then $K_{a,b}$ is L-colorable.*

Proof. Let $L_a = \{A_1, A_2, \ldots, A_a\}$ and $L_b = \{B_1, B_2, \ldots, B_b\}$. Since $a + b \leq 18$ and $a \geq 5$, we obtain $b \leq 13$. Since each color appears in at most two lists in L_b, we have $\mathcal{F} \not\subset L_b$. We can apply Strategy E if a color appears in more than $a - 5$ lists. Suppose that a color appears in exactly $a-5$ lists. Without loss of generality, assume $1 \in A_1, A_2, \ldots, A_{a-5}$. Then label the $a - 5$ lists by color 1. For the worst case, assume that color 1 is in two list in L_b, say B_1, B_2. Next, consider the remaining vertices which form $K_{5,b}$. Let L' be the list assignment of $K_{5,b}$ which is obtained from L by removing color 1. Then $L'_5 = \{A_{a-4}, A_{a-3}, A_{a-2}, A_{a-1}, A_a\}$ and $L'_b = \{B_1 - 1, B_2 - 1, B_3, \ldots, B_b\}$.

Case 1. *There is a color appearing in all lists in L'_5.*
Then L_a has a 2-coloring; hence, $K_{a,b}$ is L-colorable by Remark 1.

Case 2. *There is a color appearing in four lists in L'_5.*
By case 1, we may suppose that color 2 appears in exactly four lists in L'_5. Let $2 \in A_{a-4}, A_{a-3}, A_{a-2}, A_{a-1}$ and $A_a = 345$. We obtain three 3-colorings of L_a, that is, $\{1, 2, 3\}, \{1, 2, 4\}, \{1, 2, 5\}$. Since every color appears in at most two lists in L_b, at least one of the 3-colorings is not a list in L_b. Therefore, $K_{a,b}$ is L-colorable by Lemma 1.

Case 3. *There is a color appearing in three lists in L'_5.*
By Lemma 3, $K_{a,b}$ is L-colorable.

Case 4. *There is a color appearing in two lists in* L'_5.
From Case 3, we may suppose that each color appears in at most two lists in L'_5. Since color 1 appears in at most two lists in L_b, at most four colors appears in the same lists with color 1 in L_b. We apply Theorem 1. Since $18 \cdot \binom{3}{2} \leq \binom{10+1}{2}$, we may suppose that $|\bigcup_{i=a-4}^{a} A_i| \leq |\bigcup_{v \in V(K_{a,b})} L(v)| \leq 10$. Since $|A_{a-4}| + |A_{a-3}| + |A_{a-2}| + |A_{a-1}| + |A_a| = 15$ and the number of colors is at most ten, at least five colors must appear in exactly two lists in L'_5. Recall that only B_1, B_2 contain color 1. Hence, at most four colors from the five colors appear in the same lists with color 1 in L_b. Hence, we can choose the remaining color such that no list in L_b contains both color 1 and this color, namely color 2. Let $2 \in A_{a-4}, A_{a-3}$ and then we label A_{a-4}, A_{a-3} by color 2. Let L'' be the list assignment of $K_{3,b}$ which is obtained from L' by removing color 2. For the worst case, we suppose $2 \in B_3, B_4$. Hence, $L''_b = \{B_1 - 1, B_2 - 1, B_3 - 2, B_4 - 2, B_5 \ldots, B_b$ and $L''_2 = \{A_{a-2}, A_{a-1}, A_a\}$.

If color 3 appears in exactly two lists in A_{a-2}, A_{a-1}, A_a, then L''_3 has at least three 2-colorings containing color 3. Since every color appears in at most two lists in L''_b, at least one 2-coloring is not a list in L''_b. Otherwise, we suppose that A_{a-2}, A_{a-1}, A_a are mutually disjoint. To apply Strategy A, we count the number of lists of size 1, size 2 and size 3 in L''_b. We obtain that L''_b has no list of size 1, four lists of size 2 and $b-4$ lists of size 3 where $b - 4 \leq 6$. Then $|A_{a-2}| \cdot |A_{a-1}| \cdot |A_a| = 3^3 > 3 \cdot 4 + (b-4)$.

Case 5. $A_{a-4}, A_{a-3}, A_{a-2}, A_{a-1}, A_a$ *are mutually disjoint.*
Then L'_b has at most three lists of size $b-2$ and two lists of size 2. Since $\prod_{i=a-4}^{a} |A_i| = 3^5 > 3^3 \cdot 2 + 3^2 \cdot (b-2)$, $K_{a,b}$ is L-colorable by Strategy A for L'_5.

Strategies A,B,C,D,E and F shows that there exists a coloring of L_a such that every vertex in L_b still has available colors. It is called *Strategy A(B,C,D,E, or F) for* L_a. However, we can exchange the role between L_a and L_b for a list assignment L of $K_{a,b}$ and we call *Strategy A(B,C,D,E, or F, respectively) for* L_b.

3 On 3-Choosability of Complete Bipartite Graphs with Fourteen Vertices

Hanson, MacGillivray and Toft [4] stated that every complete bipartite graph with thirteen vertices is 3-choosable. Furthermore, Fitzpatrick and MacGillivray [3] proved that every complete bipartite graph with fourteen vertices except $K_{7,7}$ is 3-choosable. Moreover, there is the unique list assignment up to renaming the colors such that $K_{7,7}$ is not L-colorable.

Here we give a shorter proof of Fitzpatrick and MacGillivray's result by using our strategies from the previous section.

Lemma 4. *The complete bipartite graph* $K_{3,b}$ *is 3-choosable if and only if* $b \leq 26$.

Proof. Let L be the 3-list assignment of $K_{3,27}$ defined by $L_3 = \{123, 456, 789\}$ and $L_{27} = \{\{a, b, c\} | a \in \{1, 2, 3\}, b \in \{4, 5, 6\}, c \in \{7, 8, 9\}\}$. Notice that every coloring of L_3 is a list in L_{27}. By Lemma 1, $K_{3,27}$ is not L-colorable.

To prove $K_{3,26}$ is 3-choosable, let L be a 3-list assignment of $K_{3,26}$. If some lists in L_3 have a common color, $K_{3,26}$ is immediately L-colorable by Strategy B for L_3. Suppose that all lists in L_3 have no common color. We apply Strategy A for L_3 by counting the number of 3-colorings of L_3 and the number of lists of size 1, size 2 and size 3 in L_{26}. We see that the number of 3-colorings of L_3 is 27. Since L_{26} has only 26 lists of size 3, at least one of those 3-colorings is not a list in L_{26}. Hence, we can use such 3-coloring to color L_3 while every vertex in L_{26} still has an available color.

Lemma 5. *The complete bipartite graph $K_{4,10}$ is 3-choosable.*

Proof. Let L be a 3-list assignment of $K_{4,10}$. Let r_4 and r_{10} be the maximum number of lists in L_4 and L_{10}, respectively, containing a common color. Note that $r_4 \leq 4$ and $r_{10} \leq 10$.
Case 1. $r_4 = 3, 4$ or $r_{10} = 9, 10$; apply Strategy B for L_4 or Strategy B for L_{10}, respectively.
Case 2. $r_4 = 2$ and $r_{10} \leq 8$; apply Strategy C for L_4.
Case 3. $r_4 = 1$ and $r_{10} \leq 8$; apply Strategy A for L_4. Notice that $\prod_{i=1}^{4} |A_i| = 3^4 > 3 \cdot 10 = 3^{4-3} n_3$.

Lemma 6. *The complete bipartite graph $K_{5,9}$ is 3-choosable.*

Proof. Let L be a 3-list assignment of $K_{5,9}$. Let r_5 and r_9 be the maximum number of lists in L_5 and L_9, respectively, containing a common color. Then $r_5 \leq 5$ and $r_9 \leq 9$.
Case 1. $r_5 = 4, 5$ or $r_9 = 8, 9$; apply Strategy B for L_5 or Strategy B for L_9, respectively.
Case 2. $r_5 = 3$ and $r_9 \leq 7$; apply Strategy C for L_5.
Case 3. $r_5 \leq 2$ and $r_9 = 7$; apply Strategy C for L_9.
Case 4. $r_5 \leq 2$ and $r_9 = 6$; apply Strategy D for L_9.
Case 5. $r_5 \leq 2$ and $r_9 = 5$; apply Strategy E for L_9. Notice that $\mathcal{F} \not\subset L_5$ because L_5 contains only five lists.
Case 6. $r_5 = 2$ and $r_9 \leq 4$; apply Strategy D for L_5.
Case 7. $r_5 = 1$ and $r_9 \leq 4$; apply Strategy A for L_5. Notice that $\prod_{i=1}^{5} |A_i| = 3^5 > 3^2 \cdot 9 = 3^{5-3} n_3$.

Lemma 7. *The complete bipartite graph $K_{6,8}$ is 3-choosable.*

Proof. Let L be a 3-list assignment of $K_{6,8}$. Let r_6 and r_8 be the maximum number of lists in L_6 and L_8, respectively, containing a common color. Then $r_6 \leq 6$ and $r_8 \leq 8$.
Case 1. $r_6 = 5, 6$ or $r_8 = 7, 8$; apply Strategy B for L_6 or Strategy B for L_8, respectively.
Case 2. $r_6 = 4$ and $r_8 \leq 6$; apply Strategy C for L_6.
Case 3. $r_6 \leq 3$ and $r_8 = 6$; apply Strategy C for L_8.
Case 3. $r_6 \leq 3$ and $r_8 = 5$; apply Strategy D for L_8.
Case 4. $r_6 \leq 3$ and $r_8 = 4$; apply Strategy E for L_8. Notice that $\mathcal{F} \not\subset L_6$ because L_6 contains only six lists.

Case 5. $r_6 = 3$ and $r_8 \leq 3$; apply Strategy D for L_6.
Case 6. $r_6 = 2$ and $r_8 \leq 3$; apply Strategy E for L_6 unless $1 \in A_1, A_2$, $A_3 = 246, A_4 = 257, A_5 = 347, A_6 = 356$ and $\mathcal{F} \subset L_8$. In this forbidden cases, consider $3 \in A_5, A_6$ instead. Note that the forbidden cases consist of a colors in two lists and six colors in the union of the remaining four lists. In such case, colors $1, 2, 3, 4, 5, 6, 7$ have already appeared in two lists in L_6. Since $r_6 = 2$, A_1 contains two new colors, say color 8 and color 9. Hence, $A_1 \cup A_2 \cup A_3 \cup A_4$ contains eight colors, so it is not a forbidden case. Therefore, we can apply Strategy E for L_6 to conclude that $K_{6,8}$ is L-colorable.
Case 7. $r_6 = 1$ and $r_8 \leq 3$; apply Strategy A for L_6. Notice that $\prod_{i=1}^{6} |A_i| = 3^6 > 3^3 \cdot 8$.

Lemma 8. *Let L be a 3-list assignment of $K_{7,7}$. The complete bipartite graph $K_{7,7}$ is L-colorable unless $L = L_{\mathcal{F}}$.*

Proof. Let L be a 3-list assignment such that $\mathcal{F} \not\subset L_{7(i)}$ or $\mathcal{F} \not\subset L_{7(ii)}$. Let $r_{7(i)}$ and $r_{7(ii)}$ be the maximum number of lists in $L_{7(i)}$ and $L_{7(ii)}$, respectively, containing a common color. Then $r_{7(i)}, r_{7(ii)} \leq 7$.

Let $t = |\bigcup_{v \in V(K_{7,7})} L(v)|$. By Theorem 1, we may suppose that $t \leq 10$ because $14 \cdot 3 < \binom{10+1}{2}$. Since $\sum_{v \in L_{7(i)}} |L(v)| = 21$, we obtain $r_{7(i)} \geq 3$ by the Pigeonhole Principle. Similarly, $r_{7(ii)} \geq 3$.
Case 1. $r_{7(i)} = 6, 7$ or $r_{7(ii)} = 6, 7$; apply Strategy B for $L_{7(i)}$ or Strategy B for $L_{7(ii)}$, respectively.
Case 2. $r_{7(i)} = 5$ and $r_{7(ii)} \leq 5$; apply Strategy C for $L_{7(i)}$.
Case 3. $r_{7(i)} \leq 4$ and $r_{7(ii)} = 5$; apply Strategy C for $L_{7(ii)}$.
Case 4. $r_{7(i)} = 4$ and $r_{7(ii)} \leq 4$; apply Strategy D for $L_{7(i)}$.
Case 5. $r_{7(i)} = 3$ and $r_{7(ii)} = 4$; apply Strategy D for $L_{7(ii)}$.
Case 6. $r_{7(i)} = 3$ and $r_{7(ii)} = 3$; apply Strategy E for $L_{7(i)}$ unless $1 \in A_1, A_2, A_3$, $A_4 = 246, A_5 = 257, A_6 = 347, A_7 = 356$ and $L_{7(ii)} = \mathcal{F}$. In such case, $\{1, 2, 3\}, \{1, 4, 5\}, \{1, 6, 7\}$ are 3-colorings of $L_{7(ii)}$. One of such 3-colorings is not a list in $L_{7(i)}$ because $L_{7(i)} \neq \mathcal{F}$. Then $K_{7,7}$ is L-colorable by Lemma 1.

Lemmas 4, 5 6, 7 and 8 provide another proof of the next theorem of [3].

Theorem 2. *The complete bipartite graph with at most fourteen vertices is 3-choosable if and only if it is not $K_{7,7}$. For a 3-list assignment L, $K_{7,7}$ is L-colorable unless $L = L_{\mathcal{F}}$.*

4 On 3-Choosability of Complete Bipartite Graphs with Fifteen Vertices

We keep utilizing our strategies to extend the result in the previous section to 15 vertices. We first show that $K_{4,11}, K_{5,10}$ and $K_{6,9}$ are 3-choosable, and then we prove that for a 3-list assignment L, $K_{7,8}$ is L-colorable unless $L|_{V(K_{7,7})} = L_{\mathcal{F}}$.

Lemma 9. *The complete bipartite graph $K_{4,11}$ is 3-choosable.*

Proof. Let L be a 3-list assignment of $K_{4,11}$, and r_4 and r_{11} be the maximum number of lists in L_4 and L_{11}, respectively, containing a common color. Then $r_4 \leq 4$ and $r_{11} \leq 11$.
Case 1. $r_4 = 3, 4$ or $r_{11} = 10, 11$; apply Strategy B for L_4 or Strategy B for L_{11}, respectively.
Case 2. $r_4 \leq 2$ and $r_{11} = 9$; apply Strategy C for L_{11}.
Case 3. $r_4 = 2$ and $r_{11} \leq 8$; apply Strategy C for L_4.
Case 4. $r_4 = 1$ and $r_{11} \leq 8$; apply Strategy A for L_4. Notice that $\prod_{i=1}^{4} |A_i| = 3^4 > 3 \cdot 11 = 3^{4-3} n_3$.

Lemma 10. *The complete bipartite graph $K_{5,10}$ is 3-choosable.*

Proof. Let L be a 3-list assignment of $K_{5,10}$, and r_5 and r_{10} be the maximum number of lists in L_5 and L_{10}, respectively, containing a common color. Then $r_5 \leq 5$ and $r_{10} \leq 10$.
Case 1. $r_5 = 4, 5$ or $r_{10} = 9, 10$; apply Strategy B for L_5 or Strategy B for L_{10}, respectively.
Case 2. $r_5 = 3$ and $r_{10} \leq 8$; apply Strategy C for L_5.
Case 3. $r_5 \leq 2$ and $r_{10} = 8$; apply Strategy C for L_{10}.
Case 4. $r_5 \leq 2$ and $r_{10} = 7$; apply Strategy D for L_{10}.
Case 5. $r_5 \leq 2$ and $r_{10} = 6$; apply Strategy E for L_{10}. Notice that $\mathcal{F} \not\subset L_5$ because L_5 contains only five lists.
Case 6. $r_5 \leq 2$ and $r_{10} = 5$; apply Strategy F for L_{10}.
Case 7. $r_5 = 2$ and $r_{10} \leq 4$; apply Strategy D for L_5.
Case 8. $r_5 = 1$ and $r_{10} \leq 4$; apply Strategy A for L_5. Notice that $\prod_{i=1}^{5} |A_i| = 3^5 > 3^2 \cdot 10 = 3^{5-3} n_3$.

Lemma 11. *The complete bipartite graph $K_{6,9}$ is 3-choosable.*

Proof. Let L be a 3-list assignment of $K_{6,9}$, and r_6 and r_9 be the maximum number of lists in L_6 and L_9, respectively, containing a common color. Then $r_6 \leq 6$ and $r_9 \leq 9$.
Case 1. $r_6 = 5, 6$ or $r_9 = 8, 9$; apply Strategy B for L_6 or Strategy B for L_9, respectively.
Case 2. $r_6 = 4$ and $r_9 \leq 7$; apply Strategy C for L_6.
Case 3. $r_6 \leq 3$ and $r_9 = 7$; apply Strategy C for L_9.
Case 4. $r_6 \leq 3$ and $r_9 = 6$; apply Strategy D for L_9.
Case 5. $r_6 \leq 3$ and $r_9 = 5$; apply Strategy E for L_9. Notice that $\mathcal{F} \not\subset L_6$ because L_6 contains only six lists.
Case 6. $r_6 = 3$ and $r_9 \leq 4$; apply Strategy D for L_6.
Case 7. $r_6 \leq 2$ and $r_9 = 4$; apply Strategy F for L_9.
Case 8. $r_6 = 2$ and $r_9 \leq 3$; apply Strategy E for L_6 unless $1 \in A_1, A_2$ and $A_3 = 246, A_4 = 257, A_5 = 347, A_6 = 356$. In such case, we obtain that $4, 5, 6, 7 \notin A_1, A_2$ because $r_6 = 2$. Let $A_1 = 178$. Then $3 \in A_5, A_6$ and the four remaining lists cannot rename the colors to be $246, 257, 347, 356$. Hence, we still apply Strategy D for L_6.
Case 9. $r_6 = 1$ and $r_9 \leq 3$; apply Strategy A for L_6. Notice that $\prod_{i=1}^{6} |A_i| = 3^6 > 3^3 \cdot 9 = 3^{6-3} n_3$.

Lemma 12. *Let L be a 3-list assignment of $K_{7,8}$. The complete bipartite graph $K_{7,8}$ is L-colorable unless $L|_{V(K_{7,7})} = L_{\mathcal{F}}$.*

Proof. Let L be a 3-list assignment of $K_{7,8}$ such that $\mathcal{F} \not\subset L_7$ or $\mathcal{F} \not\subset L_8$. Let r_7 and r_8 be the maximum number of lists in L_7 and L_8, respectively, containing a common color. Then $r_7 \leq 7$ and $r_8 \leq 8$.
Case 1. $r_7 = 6, 7$ or $r_8 = 7, 8$; apply Strategy B for L_7 or Strategy B for L_8, respectively.
Case 2. $r_7 = 5$ and $r_8 \leq 6$; apply Strategy C for L_7.
Case 3. $r_7 \leq 4$ and $r_8 = 6$; apply Strategy C for L_8.
Case 4. $r_7 \leq 4$ and $r_8 = 5$; apply Strategy D for L_8.
Case 5. $r_7 = 4$ and $r_8 \leq 4$; apply Strategy D for L_7.
Case 6. $r_7 \leq 3$ and $r_8 = 4$; apply Strategy E for L_8 unless $1 \in B_1, B_2, B_3, B_4$, $B_5 = 246, B_6 = 257, B_7 = 347, B_8 = 356$ and $L_7 = \mathcal{F}$. Since $L_7 = \mathcal{F}$, $\{1,2,3\}, \{1,4,5\}$ and $\{1,6,7\}$ are 3-colorings of L_7. Since $\mathcal{F} \not\subset L_8$, one of such 3-colorings is not a list in L_8. Hence $K_{7,8}$ is L-colorable by Lemma 1,
Case 7. $r_7 = 3$ and $r_8 \leq 3$; apply Strategy E for L_7 unless $1 \in A_1, A_2, A_3$, $A_4 = 246, A_5 = 257, A_6 = 347, A_7 = 356$ and $\mathcal{F} \subset L_8$. In such case, let $B_1 = 123, B_2 = 145, B_3 = 146, B_4 = 246, B_5 = 257, B_6 = 347, B_7 = 356$. Suppose that $B_8 = 89A$ because $r_8 \leq 3$ and color 1 to color 7 are appears in three lists in $B_1, B_2, \ldots, B_7$. Since $L_7 \neq \mathcal{F}$, we obtain that at least one of 123, 145, 167 is not a list in L_7. Suppose that $123 \notin L_7$. Then we label $B_1, B_2, \ldots, B_7$ by color 1, color 2 and color 3. For the worst case, suppose the $2 \in A_1$ and $3 \in B_2$. Then then remaining vertices can be easily labeled. Then $K_{7,8}$ is L-colorable by Lemma 1.
Case 8. $r_7 \leq 2$ and $r_8 = 3$; apply Strategy F for L_8.
Case 9. $r_7 = 2$ and $r_8 \leq 2$; apply Strategy F for L_7.
Case 10. $r_7 = 1$ and $r_8 \leq 2$; apply Strategy A for L_7. Notice that $\prod_{i=1}^{7} |A_i| = 3^7 \geq 3^4 \cdot 8 = 3^{7-3} n_3$.

Lemmas 4, 9, 10, 11 and 12 provide us the following main theorem.

Theorem 3. *A complete bipartite graph with fifteen vertices is 3-choosable if and only if it is not $K_{7,8}$. For a 3-list assignment L, $K_{7,8}$ is L-colorable unless $L|_{V(K_{7,7})} = L_{\mathcal{F}}$.*

References

1. Charoenpanitseri, W., Punnim, N., Uiyyasathian, C.: On (k, t)−choosability of graphs. Ars. Combin. 99, 321–333 (2011)
2. Erdős, P., Rubin, A., Taylor, H.: Choosability in graphs. Congr. Num. 26, 125–157 (1979)
3. Fitzpatrick, S.L., MacGillivray, G.: Non 3-choosable bipartite graphs and the Fano plane. Ars. Combin. 76, 113–127 (2005)
4. Hanson, D., MacGillivray, G., Toft, B.: Choosability of bipartite graphs. Ars. Combin. 44, 183–192 (1996)

5. Lam, P.C.B., Shiu, W.C., Song, Z.M.: The 3-choosability of plane graphs of girth 4. Discrete Math. 294, 297–301 (2005)
6. Thomassen, C.: Every planar graph is 5-choosable. J. Combin. Theory B 62, 180–181 (1994)
7. Thomassen, C.: 3-list-coloring planar graphs of girth 5. J. Combin. Theory B 64, 101–107 (1995)
8. Vizing, V.G.: Vertex colorings with given colors. Metody Diskret. Analiz. 29, 3–10 (1976) (in Russian)
9. West, D.B.: Introduction to Graph Theory. Prentice Hall, New Jersey (2001)
10. Zhang, H.: On 3-choosability of plane graphs without 5-, 8- and 9-cycles. J. Lanzhou Univ. Nat. Sci. 41, 93–97 (2005)
11. Zhang, H., Xu, B.: On 3-choosability of plane graphs without 6-, 7- and 9-cycles. Appl. Math., Ser. B 19, 109–115 (2004)
12. Zhang, H., Xu, B., Sun, Z.: Every plane graph with girth at least 4 without 8- and 9-circuits is 3-choosable. Ars. Combin. 80, 247–257 (2006)
13. Zhu, X., Lianying, M., Wang, C.: On 3-choosability of plane graphs without 3-, 8- and 9-cycles, Australas. J. Comb. 38, 249–254 (2007)

Edge-disjoint Decompositions of Complete Multipartite Graphs into Gregarious Long Cycles*

Jung Rae Cho[1], Jeongmi Park[1], and Yoshio Sano[2]

[1] Department of Mathematics, Pusan National University, Busan 609-735, Korea
{jungcho,jm1015}@pusan.ac.kr

[2] Division of Information Engineering, Faculty of Engineering, Information and Systems, University of Tsukuba, Ibaraki 305-8573, Japan
sano@cs.tsukuba.ac.jp

Abstract. The notion of gregarious cycles in complete multipartite graphs was introduced by Billington and Hoffman in 2003 and was modified later by Billington, Hoffman, and Rodger and by Billington, Smith, and Hoffman.

In this paper, we propose a new definition of gregarious cycles in complete multipartite graphs which generalizes all of the three definitions. With our definition, we can consider gregarious cycles of long length in complete multipartite graphs, and we show some results on the existence of edge-disjoint decompositions of complete multipartite graphs into gregarious long cycles.

1 Introduction

A (simple) *graph* is a pair $G = (V, E)$ of a nonempty finite set V and a family E of 2-element subsets of V. Each element v in V is called a *vertex* of the graph G, and each element $\{u, v\}$ in E is called an *edge* of G. We denote the vertex set of G and the edge set of G by $V(G)$ and by $E(G)$, respectively. For a positive integer n, the *complete graph* of order n is the graph K_n such that $|V(K_n)| = n$ and $E(K_n) = \{\{u, v\} \mid u, v \in V(K_n), u \neq v\}$. For two positive integers n and m, the *complete multipartite graph* with n partite sets of the same size m is the graph $K_{n(m)}$ defined by $V(K_{n(m)}) := V_1 \cup \cdots \cup V_n$, where $V_1, \ldots, V_n$ are pairwise disjoint sets of the same size m, and $E(K_{n(m)}) := \{\{u, v\} \mid u \in V_i, v \in V_j, i, j \in \{1, \ldots, n\}, i \neq j\}$. We refer $V_1, \ldots, V_n$ as the *partite sets* of $K_{n(m)}$. Note that $K_{n(1)} = K_n$ and that $|V(K_{n(m)})| = nm$ and $|E(K_{n(m)})| = \frac{1}{2}n(n-1)m^2$. For a positive integer k, a *k-path* in a graph G is a sequence $(v_1, v_2, \ldots, v_k)$ of k distinct vertices of G such that $\{v_i, v_{i+1}\}$ is an edge of G for each $i = 1, 2, \ldots, k-1$. For an integer $k \geq 3$, a *k-cycle* in a graph G is a sequence $(v_1, v_2, \ldots, v_k, v_{k+1})$ of vertices of G such that $v_1, v_2, \ldots, v_k$ are distinct, $v_{k+1} = v_1$, and $\{v_i, v_{i+1}\}$ is an edge of G for each $i = 1, 2, \ldots, k$. If a vertex

* This research was supported for two years by Pusan National University Research Grant.

J. Akiyama, M. Kano, and T. Sakai (Eds.): TJJCCGG 2012, LNCS 8296, pp. 57–63, 2013.

sequence $(v_1, v_2, \ldots, v_k, v_1)$ is a k-cycle, then we denote it by $[v_1, v_2, \ldots, v_k, v_1]$. Let $k \geq 3$ be an integer. A *k-cycle decomposition* (resp. a *k-path decomposition*) of $K_{n(m)}$ is a family $\mathcal{D} = \{G_1, G_2, \ldots, G_t\}$ of k-cycles (resp. k-paths) in $K_{n(m)}$ such that $E(G_1) \cup \cdots \cup E(G_t) = E(K_{n(m)})$ and $E(G_i) \cap E(G_j) = \emptyset$ for distinct i and j in $\{1, \ldots, t\}$. For a positive integer n, we denote the n-set $\{1, \ldots, n\}$ by $[n]$.

Edge-disjoint decompositions of graphs into cycles has been considered in a number of ways. Necessary and sufficient conditions for a complete graph of odd order or a complete graph of even order minus a 1-factor to have a decomposition into cycles of some fixed length are known (see [1], [9] and [10] as well as their references). Although much work has done for cycle decompositions of complete graphs, it seems that less attention has been paid to the same problem for complete multipartite graphs. The notion of gregarious cycles in a complete multipartite graph was introduced by Billington and Hoffman [2] in 2003 and was modified later by Billington, Hoffman, and Rodger [4] and by Billington, Smith, and Hoffman [5].

In this paper, we propose a new definition of gregarious cycles in complete multipartite graphs which generalizes all of the three definitions. By introducing our new definition, we can consider gregarious cycles of length greater than the number of the partite sets in a complete multipartite graph, and we show some results on edge-disjoint decompositions of complete multipartite graphs into gregarious long cycles. This paper is organized as follows: In Section 2, we first recall the three definitions of gregarious cycles in complete multipartite graphs, and then we introduce a new definition. In Section 3, we give some results on the existence of decompositions of complete multipartite graphs into gregarious long cycles. Section 4 gives some remarks.

2 Gregarious Cycles

In 2003, Billington and Hoffman [2] introduced the notion of gregarious cycles as follows:

Definition 1 ([2]). *A 4-cycle C in a complete tripartite graph $K_{r,s,t}$ with partite sets V_1, V_2, V_3 is called* gregarious *if $|V(C) \cap V_i| \geq 1$ for any $i \in \{1, 2, 3\}$.*

They gave a necessary and sufficient condition for r, s, and t such that the complete tripartite graph $K_{r,s,t}$ has an edge-disjoint decomposition into gregarious 4-cycles.

Billington, Hoffman, and Rodger [4] defined gregarious cycles as follows:

Definition 2 ([4]). *An n-cycle C in a complete n-partite graph $K_{n(m)}$ with partite sets $V_1, \ldots, V_n$ is called* gregarious *if $|V(C) \cap V_i| = 1$ for any $i \in \{1, 2, \ldots, n\}$.*

They considered gregarious cycle decompositions of $K_{n(m)}$ with an additional property called resolvability and gave a characterization.

Billington, Smith, and Hoffman [5] defined gregarious cycles as follows:

Definition 3 **([5]).** *Let k, n be positive integers such that $k \leq n$. A k-cycle C in a complete n-partite graph $K_{n(m)}$ with partite sets $V_1, \ldots, V_n$ is called* gregarious *if $|V(C) \cap V_i| \leq 1$ for any $i \in \{1, 2, \ldots, n\}$.*

With adopting the last definition, the necessary and sufficient condition for the existence of a gregarious k-cycle decomposition of $K_{n(m)}$ is known for $k = 4$ [3], for $k = 5$ [11], for $k = 6$ and $k = 8$ [5], and for prime k [12].

As we see above, there are three different definitions of gregarious cycles. Now, we introduce a new definition of gregarious cycles which is a common generalization of the above definitions.

Definition 4. *Let n, m, and k be positive integers with $n \geq 2$ and $k \geq 3$. A k-cycle C in a complete n-partite graph $K_{n(m)}$ with partite sets $V_1, \ldots, V_n$ is called* gregarious *if*

$$\left\lfloor \frac{k}{n} \right\rfloor \leq |V(C) \cap V_i| \leq \left\lceil \frac{k}{n} \right\rceil$$

for any $i \in \{1, 2, \ldots, n\}$.

Remark 1. (i) If we take $k = 4$ and $n = 3$, then the above definition coincides with that of Billington and Hoffman [2].
(ii) If we take $k = n$, then the above definition coincides with that of Billington, Hoffman, and Rodger [4].
(iii) For the case where $k \leq n$, the above definition coincides with that of Billington, Smith, and Hoffman [5].
(iv) More generally, we can define as follows: A subgraph H of a multipartite graph G with partite sets $V_1, \ldots, V_n$ is said to be *gregarious* in G if

$$\left\lfloor \frac{1}{n}|V(H)| \right\rfloor \leq |V(H) \cap V_i| \leq \left\lceil \frac{1}{n}|V(H)| \right\rceil$$

for any $i \in \{1, 2, \ldots, n\}$. This is an extension of the definition given by Smith [12]. □

With our definition, we can consider gregarious cycles of length greater than the number of the partite sets in a complete multipartite graph. A *gregarious cycle decomposition* of $K_{n(m)}$ is a cycle decomposition in which all the cycles are gregarious in the sense of our definition. Here are some examples of gregarious k-cycle decompositions of $K_{n(m)}$ with $k > n$. In the following, we let $V_i := \{i_1, \ldots, i_m\}$ for each $i \in \{1, \ldots, n\}$.

Example 1. The complete multipartite graph $K_{4(2)}$ has a gregarious 6-cycle decomposition. Let

$$C_1 = [1_1, 2_1, 3_1, 4_1, 3_2, 2_2, 1_1],\ C_2 = [1_2, 2_1, 3_2, 4_2, 3_1, 2_2, 1_2],$$
$$C_3 = [2_1, 4_1, 1_1, 3_1, 1_2, 4_2, 2_1],\ C_4 = [2_2, 4_1, 1_2, 3_2, 1_1, 4_2, 2_2].$$

Then we can check that C_1, C_2, C_3, C_4 are gregarious 6-cycles in $K_{4(2)}$ and $\{C_1, C_2, C_3, C_4\}$ is a cycle decomposition of $K_{4(2)}$. □

Example 2. The complete multipartite graph $K_{4(4)}$ has a gregarious 8-cycle decomposition. Let

$$\begin{array}{ll} C_1 = [1_1, 2_1, 3_2, 4_2, 1_3, 2_3, 3_4, 4_4, 1_1], & C_7 = [1_2, 2_2, 3_1, 4_1, 1_4, 2_4, 3_3, 4_3, 1_2], \\ C_2 = [1_3, 2_1, 3_4, 4_2, 1_1, 2_3, 3_2, 4_4, 1_3], & C_8 = [1_4, 2_2, 3_3, 4_1, 1_2, 2_4, 3_1, 4_3, 1_4], \\ C_3 = [1_1, 3_1, 4_2, 2_2, 1_3, 3_3, 4_4, 2_4, 1_1], & C_9 = [1_2, 3_2, 4_1, 2_1, 1_4, 3_4, 4_3, 2_3, 1_2], \\ C_4 = [1_3, 3_1, 4_4, 2_2, 1_1, 3_3, 4_2, 2_4, 1_3], & C_{10} = [1_4, 3_2, 4_3, 2_1, 1_2, 3_4, 4_1, 2_3, 1_4], \\ C_5 = [1_1, 4_1, 2_2, 3_2, 1_3, 4_3, 2_4, 3_4, 1_1], & C_{11} = [1_2, 4_2, 2_1, 3_1, 1_4, 4_4, 2_3, 3_3, 1_2], \\ C_6 = [1_3, 4_1, 2_4, 3_2, 1_1, 4_3, 2_2, 3_4, 1_3], & C_{12} = [1_4, 4_2, 2_3, 3_1, 1_2, 4_4, 2_1, 3_3, 1_4]. \end{array}$$

Then we can check that $C_1, \ldots, C_{12}$ are gregarious 8-cycles and $\{C_1, \ldots, C_{12}\}$ is a cycle decomposition of $K_{4(4)}$. □

The following proposition gives a necessary condition for $K_{n(m)}$ to have a gregarious k-cycle decomposition where k may be larger than n.

Proposition 1. *Let n, m, k be positive integers such that $k \geq 3$. If there exists a gregarious k-cycle decomposition of $K_{n(m)}$, then $(n-1)m$ is even and $\frac{1}{2}n(n-1)m^2$ is divided by k.*

Proof. Since the degree of each vertex should be even, $(n-1)m$ is even. Since the number of edges of $K_{n(m)}$ should be divided by the length of the cycles, $\frac{1}{2}n(n-1)m^2$ is divided by k. □

3 Results

In this section, we show some results on edge-disjoint decompositions of complete multipartite graphs into gregarious cycles of long length. In particular, we consider the cases where the length k of cycles is equal to $2n-2$ or $2n$, where n is the number of partite sets.

3.1 Gregarious $(2n-2)$-cycle Decompositions of $K_{n(m)}$

In this subsection, we prove the following theorem.

Theorem 1. *Let n and m be positive integers with $n > 2$. If n and m are even, then the complete multipartite graph $K_{n(m)}$ has a gregarious $(2n-2)$-cycle decomposition.*

Lemma 1. *The complete graph K_n has an n-path decomposition if and only if n is even.*

Proof. Let n be an even positive integer. Then, since $n+1$ is odd, K_{n+1} has an $(n+1)$-cycle decomposition $\mathcal{D}$ (cf. [1]). Fix a vertex v^* of K_{n+1}. For each $(n+1)$-cycle $C \in \mathcal{D}$, $C - v^*$ is an n-path. Then we can easily check that the family $\{C - v^* \mid C \in \mathcal{D}\}$ is an n-path decomposition of $K_{n+1} - v^* \cong K_n$. Conversely, if K_n has an n-path decomposition, then $|E(K_n)| = \frac{1}{2}n(n-1)$ must be divided by $(n-1)$ and thus n is even. □

Lemma 2. *Let n and m be positive integers with $n > 2$. If n is even, then the complete multipartite graph $K_{n(m)}$ has a gregarious n-path decomposition.*

Proof. Since n is even, it follows from Lemma 1 that there exists an n-path decomposition $\mathcal{D}$ of the complete graph K_n with $V(K_n) = \{1, 2, \ldots, n\}$. Note that $K_n = K_{n(1)}$ and that n-paths in $K_{n(1)}$ are gregarious n-paths. Let $V_i = \{i_1, \ldots, i_m\}$ ($i \in [n]$) denote the partite sets of $K_{n(m)}$. For each n-path $P \in \mathcal{D}$ of the form $P = (\sigma(1), \sigma(2), \ldots, \sigma(n-1), \sigma(n))$, where σ is a permutation of $[n]$, we define m^2 gregarious n-paths in $K_{n(m)}$ by

$$Q_{ij}^P := (\sigma(1)_i, \sigma(2)_j, \sigma(3)_i, \sigma(4)_j, \ldots, \sigma(n-1)_i, \sigma(n)_j) \quad ((i,j) \in [m] \times [m]).$$

Then we can check that Q_{ij}^P is gregarious and that $\mathcal{D}^* := \{Q_{ij}^P \mid P \in \mathcal{D}, (i,j) \in [m] \times [m]\}$ gives an edge-disjoint path decomposition of $K_{n(m)}$. Hence the complete multipartite graph $K_{n(m)}$ has a gregarious n-path decomposition. □

Proof (Proof of Theorem 1). Since m is even, $l := \frac{1}{2}m$ is an integer. Since n is even, it follows from Lemma 2 that there exists a gregarious n-path decomposition $\mathcal{D}$ of the complete multipartite graph $K_{n(l)}$. For each gregarious n-path $P \in \mathcal{D}$, P has the form

$$P = (\sigma(1)_{f(1)}, \ldots, \sigma(n)_{f(n)})$$

where σ is a permutation of $[n]$ and f is a map from $[n]$ to $[l]$. For simplicity, we denote $\sigma(i)_{f(i)}$ by x_i. Let $x'_i := \sigma(i)_{f(i)+l}$. For each gregarious n-path $P = (x_1, x_2, \ldots, x_n) \in \mathcal{D}$, we define two gregarious $(2n-2)$-cycles C_+^P and C_-^P in $K_{n(2l)}$ by

$$C_+^P := [x_1, x_2, x_3, \ldots, x_{n-2}, x_{n-1}, x_n, x'_{n-1}, x'_{n-2}, x'_{n-3}, \ldots, x'_3, x'_2, x_1],$$
$$C_-^P := [x'_1, x_2, x'_3, \ldots, x_{n-2}, x'_{n-1}, x'_n, x_{n-1}, x'_{n-2}, x_{n-3}, \ldots, x_3, x'_2, x'_1].$$

Then we can check that $\mathcal{D}^* := \{C_\epsilon^P \mid P \in \mathcal{D}, \epsilon \in \{+,-\}\}$ gives a gregarious $(2n-2)$-cycle decomposition of $K_{n(2l)} = K_{n(m)}$. Hence the theorem holds. □

3.2 Gregarious $2n$-cycle Decompositions of $K_{n(m)}$

In this subsection, we show the following the following theorem.

Theorem 2. *Let n and m be positive integers with $n > 2$. If n is even and m is a multiple of 4, then the complete multipartite graph $K_{n(m)}$ has a gregarious $2n$-cycle decomposition.*

To prove this theorem, we use the following result by Billington, Hoffman, and Rodger.

Lemma 3 ([4, Corollary 2.2]). *Let n and m be positive integers with $n > 2$. If n and m are even, then the complete multipartite graph $K_{n(m)}$ has a gregarious n-cycle decomposition.* □

Proof (Proof of Theorem 2). Since m is a multiple of 4, $l := \frac{1}{2}m$ is an even integer. Since n and l are even, it follows from Lemma 3 that there exists a gregarious n-cycle decomposition $\mathcal{D}$ of the complete multipartite graph $K_{n(l)}$. For each gregarious n-cycle $C \in \mathcal{D}$, C has the form $C = [\sigma(1)_{f(1)}, \dots, \sigma(n)_{f(n)}, \sigma(1)_{f(1)}]$ where σ is a permutation of $[n]$ and f is a map from $[n]$ to $[l]$. For simplicity, we denote $\sigma(i)_{f(i)}$ by x_i. Let $x'_i := \sigma(i)_{f(i)+l}$. For each gregarious n-cycle $C = [x_1, \dots, x_n, x_1] \in \mathcal{D}$, we define two gregarious $2n$-cycles O^C_+ and O^C_- in $K_{n(2l)}$ by

$$O^C_+ := [x_1, x_2, x_3, \dots, x_{n-2}, x_{n-1}, x_n, x'_1, x'_2, x'_3, \dots, x'_{n-2}, x'_{n-1}, x'_n, x_1],$$
$$O^C_- := [x'_1, x_2, x'_3, \dots, x_{n-2}, x'_{n-1}, x_n, x_1, x'_2, x_3, \dots, x'_{n-2}, x_{n-1}, x'_n, x'_1].$$

Then we can check that $\mathcal{D}^* := \{O^C_\epsilon \mid C \in \mathcal{D}, \epsilon \in \{+, -\}\}$ gives a gregarious $2n$-cycle decomposition of $K_{n(2l)} = K_{n(m)}$. Hence the theorem holds. □

4 Concluding Remarks

In this paper, we introduced a new definition of gregarious cycles in complete multipartite graphs as a common generalization of three different definitions, and we show some results on the existence of edge-disjoint decompositions of complete multipartite graphs into gregarious long cycles. We left the existence problems of gregarious cycle decompositions of complete multipartite graphs in other various cases for further study, especially the case where the length of cycles is greater than the number of partite sets. It would be also interesting to consider gregarious long cycle decompositions having some additional conditions such as resolvable decompositions (see [4]) and circulant decompositions (see [6], [7], [8]).

References

1. Alspach, B., Gavlas, H.: Cycle decompositions of K_n and $K_n - I$. Journal of Combinatorial Theory, Series B 81, 77–99 (2001)
2. Billington, E.J., Hoffman, D.G.: Decomposition of complete tripartite graphs into gregarious 4-cycles. Discrete Mathematics 261, 87–111 (2003)
3. Billington, E.J., Hoffman, D.G.: Equipartite and almost-equipartite gregarious 4-cycle systems. Discrete Mathematics 308, 696–714 (2008)
4. Billington, E.J., Hoffman, D.G., Rodger, C.A.: Resolvable gregarious cycle decompositions of complete equipartite graphs. Discrete Mathematics 308, 2844–2853 (2008)
5. Billington, E.J., Smith, B.R., Hoffman, D.G.: Equipartite gregarious 6- and 8-cycle systems. Discrete Mathematics 307, 1659–1667 (2007)
6. Cho, J.R.: A note on decomposition of complete equipartite graphs into gregarious 6-cycles. Bulletin of the Korean Mathematical Society 44, 709–719 (2007)
7. Cho, J.R., Gould, R.J.: Decompositions of complete multipartite graphs into gregarious 6-cycles using complete differences. Journal of the Korean Mathematical Society 45, 1623–1634 (2008)

8. Kim, E.K., Cho, Y.M., Cho, J.R.: A difference set method for circulant decompositions of complete partite graphs into gregarious 4-cycles. East Asian Mathematical Journal 26, 655–670 (2010)
9. Šajna, M.: On decomposiing $K_n - I$ into cycles of a fixed odd length. Descrete Mathematics 244, 435–444 (2002)
10. Šajna, M.: Cycle decompositions III: complete graphs and fixed length cycles. Journal of Combinatorial Designs 10, 27–78 (2002)
11. Smith, B.R.: Equipartite gregarious 5-cycle systems and other results. Graphs and Combinatorics 23, 691–711 (2007)
12. Smith, B.R.: Some gregarious cycle decompositions of complete equipartite graphs. The Electronic Journal of Combinatorics 16(1), Research Paper 135, 17 (2009)

Affine Classes of 3-Dimensional Parallelohedra - Their Parametrization -

Nikolai Dolbilin[1,⋆], Jin-ichi Itoh[2,⋆⋆], and Chie Nara[3,⋆⋆⋆]

[1] Steklov Institute of Mathematics, Russian Academy of Science, ul. Gubkina 8, Moscow, 119991, Russia
dolbilin@mi.ras.ru

[2] Faculty of Education, Kumamoto University, Kumamoto, 860-8555, Japan
j-itoh@kumamoto-u.ac.jp

[3] Liberal Arts Education Center, Aso Campus, Tokai University, Aso, Kumamoto, 869-1404, Japan
cnara@ktmail.tokai-u.jp

Abstract. In addition to the well-known classification of 3-dimensional parallelohedra we describe this important class of polytopes classified by the affine equivalence relation and parametrize representatives of their equivalent classes.

1 Introduction

For each dimension, parallelohedra constitute a very important class of Euclidean polyhedra that have important applications in geometry, especially in geometry of numbers, combinatorial geometry, and in some other fields of mathematics. Three-dimensional parallelohedra play a significant role in geometric crystallography. The concept and the term of a paralleloherdon were introduced by the Russian eminent crystallographer E.S.Fedorov ([1]).

A *d-parallelohedron* is defined as a polyhedron whose parallel copies tile the space R^d in a face-to-face manner. Classical theorems by H. Minkowski [2] and B. A. Venkov [3] are equivalent to the following criterion:

Theorem 1. *([3,4]). A d-dimensional convex bounded polyhedron is a parallelohedron if and only if*
(i) P is centrally symmetric;
(ii) all its faces are centrally symmetric;
(iii) the projection of P along each of its $(d-2)$-faces is either a parallelogram or a centrally symmetric hexagon.

E. S. Fedorov [1] determined all five combinatorial types of convex 3-parallelohedra.

⋆ Supported by the RF goverment grant N 11.G34.31.0053 and the RFBR grant 11-01-00633-a.
⋆⋆ Supported by Grant-in Aid for Scientific Research (No. 23540098), JSPS.
⋆⋆⋆ Supported by Grant-in Aid for Scientific Research (No. 23540160), JSPS.

J. Akiyama, M. Kano, and T. Sakai (Eds.): TJJCCGG 2012, LNCS 8296, pp. 64–72, 2013.

Theorem 2. *([1]) There are five combinatorial types of convex parallelohedra in R^3: the cube, the right hexagonal prism, the rhombic dodecahedron, the elongated dodecahedron, and the truncated octahedron.*

A well-developed, algorithmical theory of a very important subclass of parallelohedra had been elaborated by G.Voronoi [4]. This subclass consists of those parallelohedra which can be represented as Dirichlet-Voronoi domains of points in a integer point lattice. Now such parallelohedra are called *Voronoi parallelohedra.* Not every parallelohedron is a Voronoi parallelohedron. So, for instance in a plane every 2-dimensional parallelohedron (syn. *parallelogon*) is either a parallelogram or a centrally symmetric hexagon. However, a parallelogon is a Voronoi 2-dimensional parallelohedron if and only if it is either a rectangle or a centrally symmetric hexagon inscribed into a circle.

Voronoi introduced a notion of a *primitive parallelohedron* as a parallelohedron to tile a space in such a way that each vertex belongs to the least possible number (for a given dimension d) of tiling cells, namely, $d+1$. In a space of dimension 2 or 3 there is the only combinatorial type of primitive parallelohedra (if $d=2$ it is the hexagon but not the parallelogram, if $d=3$, it is the truncated octahedron only). If $d=4$ or 5, there are 3 or 222 combinatorial types of primitive parallelohedra, respectively. Voronoi proved that *every primitive parallelohedron is affine equivalent to some Voronoi parallelohedron* and suggested a conjecture: *For any parallelohedron there exists an affine equivalent* (for brevity, *a-equivalent*) *Voronoi parallelohedron.* Regardless of serious efforts and significant progresses this centennial conjecture on the *existence* of the a-equivalent Voronoi parallelohedron still remains unsolved. Among recent results we select out so-called *uniqueness* theorems. In [5] it was proved that if a parallelohedron P is primitive, then an a-equivalent Voronoi parallelohedron P' is determined uniquely up to similarity. The uniqueness theorem was proved in [6] in a very elementary way for a wider class of parallelohedra, namely for those parallelohedra whose boundary after the removal of all standard faces (see [6] for definition) remains connected.

The uniqueness theorem easily implies a surprising fact. As already said, Voronoi developed a deep theory of Voronoi parallelohedra ([4]). According to this theory, all Voronoi parallelohedra of a given primitive combinatorial type correspond to lattices which fill a so-called *Voronoi type domain* in the cone of positive quadratic forms. If, instead a Voronoi tiling by primitive parallelohedra, one considers dual Delaunay tiling (in the primitive case by simplexes), in the interior of a given Voronoi type domain *all the Delaunay tilings* are pairwise affine equivalent. The surprising fact to follow from the uniqueness theorem is that *all Voronoi tilings are pairwise affine non-equivalent,* in contrast to the uniqueness of affine classes of Delaunay tilings within the domain.

In the case of d-dimensional primitive parallelohedra the dimension of each Voronoi type domain is equal to $\frac{d(d+1)}{2}$. Thus, from the uniqueness theorem ([5,6]) it follows that the dimension of the space of affine equivalence classes is equal to $d(d+1)/2-1$. So, if $d=2$, for example, in the primitive case (the centrally symmetrical hexagon) the dimension of the space of affine equivalence classes of a primitive parallelohedron is 2. If $d = 3$ the dimension of the space of affine equivalence classes with combinatorial type of the truncated octahedron is equal to 5.

We see that the affine classification of parallelohedra turns out a delicate question relevant to the Voronoi conjecture. In this paper, we classify convex parallelohedra in R^3 by the affine equivalence relation and realize their representatives in geometric formulation. In this way we will find the dimension of the space of affine equivalence classes of all 5 different combinatorial types of parallelohedra in 3-space (Theorems 4-6).

We study on centrally symmetric hexagons in Sect. 2 and truncated octahedrons in Sect. 3. The main theorems are showed in Sect. 3 for primitive parallelohedra and in Sect. 4 for non-primitive parallelohedra. The affine equivalent classes of parallelohedra with the combinatorial type of the truncated octahedron, are parameterized by a 5-tuple $(\alpha, \beta, h, \delta, l)$ which satisfies $0 < \alpha,\ 0 < \beta \leq (\pi - \alpha)/2, 0 \leq h, 0 < \pi - \gamma < \tan^{-1}(\sin(\alpha/2)/h), 0 < \pi - \delta < \tan^{-1}(\sin(\beta/2)/h)$, and the inequalities (4) and (5) given in the section 3 (Theorem 4). The affine classes of parallelohedra with the combinatorial type of the rhombic dodecahedron are parameterized by a 3-tuple $(\alpha,\ \beta, h)$ where $0 < \alpha, 0 < \beta \leq (\pi - \alpha)/2, 0 < h$ (Theorem 5).

2 Two-Dimensional Case

We start with parallelogons, i.e. 2-dimensional parallelohedra. There are two combinatorial types of parallelogons: the quadrangle and the hexagon. Moreover, since parallelohedra are centrally symmetric, a parallelogon is either a parallelogram or a centrally symmetrical hexagon.

All parallelograms are obviously pairwise a-equivalent, i.e. belonging to one affine class. The dimension of the space of affine classes of parallelograms is zero.

Now give a centrally symmetric (c.-s.) hexagon. A c.-s. hexagon is inscribed into an ellipse. By an appropriate affine map the ellipse is transformed on a unit circle. Let O be the center of the unit circum-circle of the hexagon transformed by the affine map, and let $A_1, A_2, A_3, A_4, A_5, A_6$ be vertices of the hexagon. A centrally symmetric hexagon has three pairs of central angles symmetric each other: $A_1\widehat{O}A_2 = A_4\widehat{O}A_5$, $A_2\widehat{O}A_3 = A_5\widehat{O}A_6$, and $A_3\widehat{O}A_4 = A_6\widehat{O}A_1$. Let α, β, γ be the values of these angles. Without loss of generality we can and will consider only the following triples

$$0 < \alpha \leq \beta \leq \gamma, \text{ where } \alpha + \beta + \gamma = \pi. \tag{1}$$

Each triple α, β, γ with (1) determines a unique (up to congruence) a-equivalent c.-s. hexagon inscribed into a unit circle, and vice versa. On the other hand, two inscribed c.-s. hexagons with different triples satisfying (1) $(\alpha, \beta, \gamma) \neq (\alpha', \beta', \gamma')$ are *not* a-equivalent.

Theorem 3. *The configuration space of a-equivalence classes of centrally symmetric (c.-s.) hexagons has dimension 2 and can be parameterized by ordered triples* (α, β, γ) *provided* $0 < \alpha \leq \beta \leq \gamma$, $\alpha + \beta + \gamma = \pi$.

3 The Truncated Octahedron

For a given combinatorial type K of parallelohedra, we denote by $\mathcal{A}(K)$ the set of the affine equivalence classes of parallelohedra combinatorially equivalent to K.

For a given parallelohedron P and each $(d-2)$-face of P, there is a cycle of four or six $(d-1)$-faces by Theorem 1 (*iii*). We call this cycle a *belt* of P.

In the rest of this section, we consider P a parallelohedron with its combinatorial type of the truncated octahedron. So, P has six different belts. Each belt consists of six faces (two parallelograms and four centrally symmetric hexagons) and it has six parallel edges by Theorem 1 (*iii*).

Lemma 1. *Six centers of faces on a belt and the center of P are coplanar.*

Proof. Let the center of P be the origin O in R^3, and G_i be centers of six consecutive faces $F_i\,(1 \le i \le 6)$ of a belt of P. Since P is centrally symmetric, $\overrightarrow{OG_i} = -\overrightarrow{OG_{i+3}}$ for $1 \le i \le 3$. Since P is a parallelohedron, P tiles the space by its parallel copies in a face-to face manner. Let P_1 and P_2 be the copies of P obtained by the parallel translations along $2\overrightarrow{OG_1}$ and $2\overrightarrow{OG_2}$ respectively. Since P is primitive, the edge $F_1 \cap F_2$ belongs to exactly three parallel copies of P (including itself) in its tiling. So, P_2 is obtained by the parallel translation of P_1 along $2\overrightarrow{OG_3}$. Hence $\overrightarrow{OG_3} = \overrightarrow{OG_2} - \overrightarrow{OG_1}$. Therefore, $G_i\,(1 \le i \le 6)$ and the origin are coplanar.

Now we fix a belt of P, and define a reduced parallelohedron P_r of P corresponding to the belt, which is described in R^3 with the origin O as the center of P.

Step 1. We can assume all centers of the faces of the belt is on the xy-plane by Lemma 1 and the center of P is the origin. The orthogonal projection of P to the xy-plane is a centrally symmetric hexagon by Theorem 1 (iii).

Step 2. There is an affine transformation which satisfies the following conditions:
i) the c.-s. hexagon in the xy-plane is mapped to a c.-s. hexagon inscribed in the unit circle with center O in xy-plane, by Theorem 3, and
ii) the parallel six edges of the belt are mapped to edges with unit length, which are parallel to the z-axis.

We denote such transformation by $f_a = f_{a,\mathcal{B}}$ which depends on the belt $\mathcal{B}$. We call the image P_r of P by f_a a *reduced parallelohedron* of P. Now we show that P_r is uniquely determined by the following five parameters.

Definition of Parameters. Let $\pm F_1$ be two parallelograms and $\pm F_i\,(i = 2, 3)$ be four hexagons in the belt of P_r, where F_1, F_2 and F_3 are consecutive in order. Let α (resp. β) be $\angle A_1OA_2$ (resp. $\angle A_2OA_3$) where the line segment A_1A_2 (resp. A_2A_3) is the projection of F_1 (resp. F_2) to the xy-plane. We can assume $\beta \le (\pi - \alpha)/2$ by considering $-F_3$ instead of F_2 if necessary.

Let B_1, B_2, B_3, B_4, B_5 and B_6 be consecutive vertices of F_2, where the line segment B_1B_2 is the common edge of F_1 and F_2 and the z-coordinate of B_2 is greater than the one of B_1. Notice that we can assume the z-coordinate of the midpoint of the edge B_1B_2, denoted by h, satisfies $h \geq 0$, by considering $-F_1$ and $-F_2$ instead of F_1 and F_2 if necessary. Denote $\angle B_1B_2B_3$ and $\angle B_3B_4B_5$ by γ and δ respectively (see Fig. 2).

Let C_1 and C_2 be vertices of F_1 so that $F_1 = B_1B_2C_2C_1$. By $|A_1A_2| = 2\sin(\alpha/2)$, $\angle B_1B_2C_2 = \tan^{-1}(|A_1A_2|/2h) = \tan^{-1}(\sin(\alpha/2)/h)$, where $|XY|$ means the Euclidean distance of X, $Y \in R^3$. Since P is convex, $\triangle B_2B_3C_2$ is upper than the parallelogram $B_2C_2(-B_1)(-C_1)$, and so $\gamma > \pi - \tan^{-1}(\sin(\alpha/2)/h)$. Since F_2 is convex, $\gamma < \pi$. Hence

$$0 < \pi - \gamma < \tan^{-1}(\sin(\alpha/2)/h). \tag{2}$$

Since $\angle B_2B_4B_5 < \delta < \pi$ and $\angle B_2B_4B_5 = \pi - \tan^{-1}(\sin(\beta/2)/h)$, δ satisfies

$$0 < \pi - \delta < \tan^{-1}(\sin(\beta/2)/h). \tag{3}$$

For a point Q and a set S in R^3, we denote by $-Q$ the symmetric point of Q about the origin, and by $-S$ the set $\{-Q : Q \in S\}$. For three points P_1, P_2 and P_3 in R^3 which are not collinear, we denote by $\Pi(x, y, z;\ P_1, P_2, P_3) = 0$ the equation of the plane including those three points and by $\Pi(Q;\ P_1, P_2, P_3)$ the value $\Pi(q_x, q_y, q_z;\ P_1, P_2, P_3)$ for a point $Q = (q_x, q_y, q_z)$.

Since the plane including $\triangle B_2B_3C_2$ (resp. $\triangle(-B_1)(-B_6)(-C_1)$) does not intersect with the edge $(-B_1)(-B_6)$ (resp. $(B_2)(B_3)$),

$$\Pi(A_2;\ B_2, B_3, C_2) \cdot \Pi(-B_6;\ B_2, B_3, C_2) > 0 \tag{4}$$

and

$$\Pi(A_2;\ -B_1,\ -C_1,\ -B_6) \cdot \Pi(B_3;\ -B_1,\ -C_1,\ -B_6) > 0 \tag{5}$$

hold , where

(i) $A_1A_2 \cdots A_6$ is the hexagon centrally symmetric about the origin with vertices $A_1 = (\cos\alpha, \sin\alpha, 0)$, $A_2 = (1, 0, 0)$, $A_3 = (\cos\beta,\ -\sin\beta, 0)$,

(ii) $B_1B_2 \cdots B_6$ is the hexagon centrally symmetric about the midpoint of A_2A_3 with vertices $B_1 = (1, 0,\ -1/2 + h)$, $B_2 = (1, 0,\ 1/2 + h)$ and the point B_3 determined by $\angle B_1B_2B_3 = \gamma$ and $\angle B_3B_4B_5 = \delta$, and

(iii) C_1 and C_2 are the points symmetric to B_2 and B_1 respectively about the midpoint of A_1A_2 (see Figs. 1, 2 and 4).

We call such 5-tuple $(\alpha, \beta, h, \delta, l)$ a *parameterization of* P_r. We show that for each 5-tuple satisfying the above conditions, there exists a unique parallelohedron with the given parametrization and the combinatorial type of the truncated octahedron.

Theorem 4. *The affine equivalent classes* $\mathcal{A}(K)$ *of parallelohedra with the combinatorial type* K *of the truncated octahedron are parameterized by a 5-tuple* $(\alpha, \beta, h, \delta, l)$ *which satisfies the following:*

$$0 < \alpha,\ 0 < \beta \leq (\pi - \alpha)/2,\ 0 \leq h,$$

$$0 < \pi - \gamma < \tan^{-1}(\sin(\alpha/2)/h),$$
$$0 < \pi - \delta < \tan^{-1}(\sin(\beta/2)/h),$$

and the conditions (4) and (5).

Proof. Let $(\alpha, \beta, h, \delta, l)$ be a 5-tuple satisfying all conditions in the theorem.

Step 1. Take a c.-s. hexagon $A_1A_2 \cdots A_6$ in R^3 satisfying the following conditions (1)-(4): (1) inscribed in the unit circle with the center of the origin O, (2) included in the xy-plane, (3) $\angle A_1OA_2 = \alpha$ and $\angle A_2OA_3 = \beta$, and (4) the point A_2 is in the positive x-axis (see the left figure in Fig. 1).

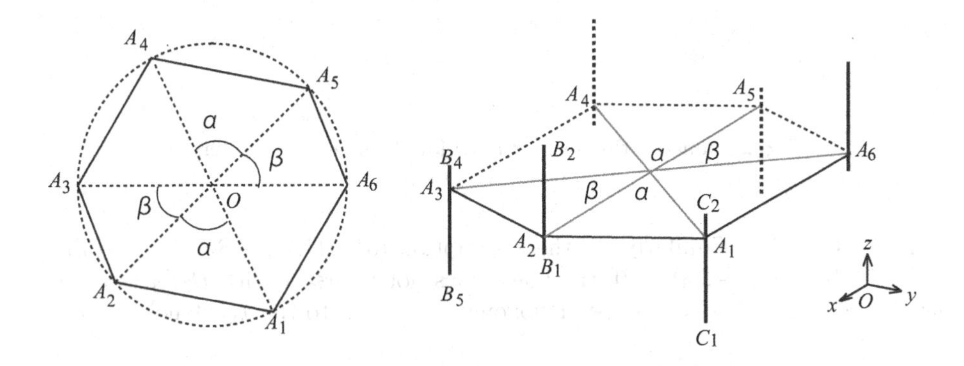

Fig. 1. Steps to obtain a truncated octahedron

Step 2. Let e_1 be the line segment with unit length included in the line passing through A_1, and orthogonal to the xy-plane, whose midpoint has the z-coordinate $-h$. Draw five edges e_i $(i = 2, \cdots, 6)$ with unit length parallel to e_1 such that e_{i+1} is symmetric to e_i about the midpoint of A_iA_{i+1} for each $i = 1, \cdots, 6$, where e_7 means e_1. Denote by $e_1 = C_1C_2$, $e_2 = B_1B_2$, and $e_3 = B_4B_5$, where the z-coordinate of B_2 (resp. B_4) is greater than the one of B_1 (resp. B_5) (see the right figure in Fig. 1).

Step 3. Let F_1 be the parallelogram $C_1C_2B_2B_1$ spanned by e_1 and e_2. Let F_2 be a c.-s. hexagon $B_1B_2 \cdots B_6$ with angles $\angle B_1B_2B_3 = \gamma$ and $\angle B_3B_4B_5 = \delta$ (see Fig. 2).

Step 4. Let Π_1 be the plane including the edges B_2B_3 and B_2C_2 . Let Π_2 be the plane including the edges $(-C_1)(-B_1)$ of $-F_1$ and $(-B_1)(-B_6)$ of $-F_2$.

By the conditions (2) and (3), B_3 and $-B_6$ are higher than the plane including the parallelogram $B_2C_2(-B_1)(-C_1)$.

Since Π_1 and Π_2 include parallel lines B_2C_2 and $(-C_1)(-B_1)$ respectively, and cannot be parallel planes from the existence of B_3 (upper than B_2) and $-B_6$ (upper than $-B_1$), the two planes Π_1 and Π_2 intersect in a line (denoted by l) which is parallel to B_2C_2 and $(-C_1)(-B_1)$.

By the assumption (4), two points $-B_6$ and A_2 are in the same half-space divided by the equation $\Pi(x, y, z; B_2, B_3, C_2) = 0$. So, l does not intersect with the

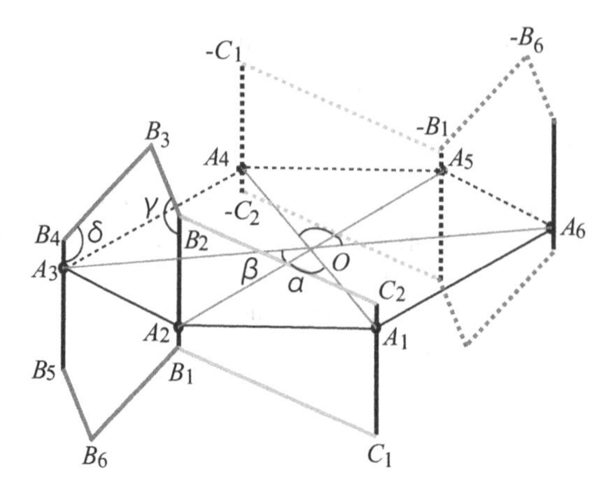

Fig. 2. Construction of four faces for the given parameters

edge $(-B_1)(-B_6)$, Similarly, by the assumption (5), $\Pi(A_2; -B_1, -C_1, -B_6) \cdot \Pi(B_3; -B_1, -C_1, -B_6) > 0$, the line l does not intersect with B_2B_3 (see the left figure in Fig. 3 which is the orthogonal projection to the xy-plane).

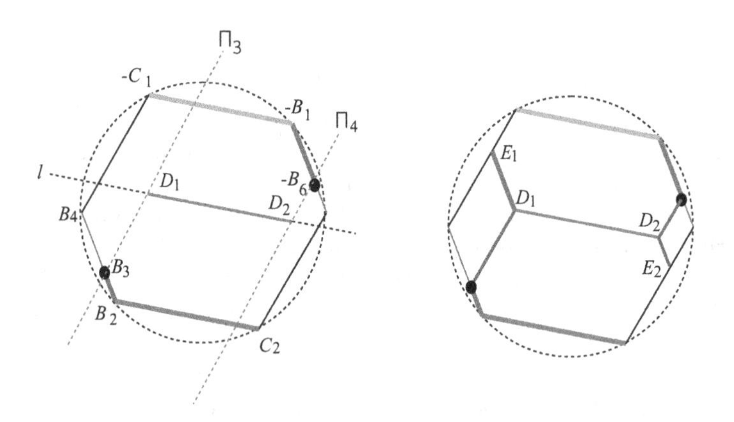

Fig. 3. The orthogonal projection to the xy-plane

Step 5. Let Π_3 (resp. Π_4) be the plane which is orthogonal to the xy-plane, parallel to A_3A_4, and which includes the point B_3 (resp. $-B_6$). Denote by D_1 (resp. D_2) the intersection point of the line l and Π_3 (resp. Π_4) (see the left figure in Fig. 3).

Step 6. Let E_1 (resp. E_2) be the point such that the line segment D_1E_1 (resp. D_2E_2) is parallel and congruent to the edge B_3B_4 (resp. B_2B_3) (see the right figure in Fig. 3). Now we obtain a belt (see the left figure in Fig. 4). By drawing

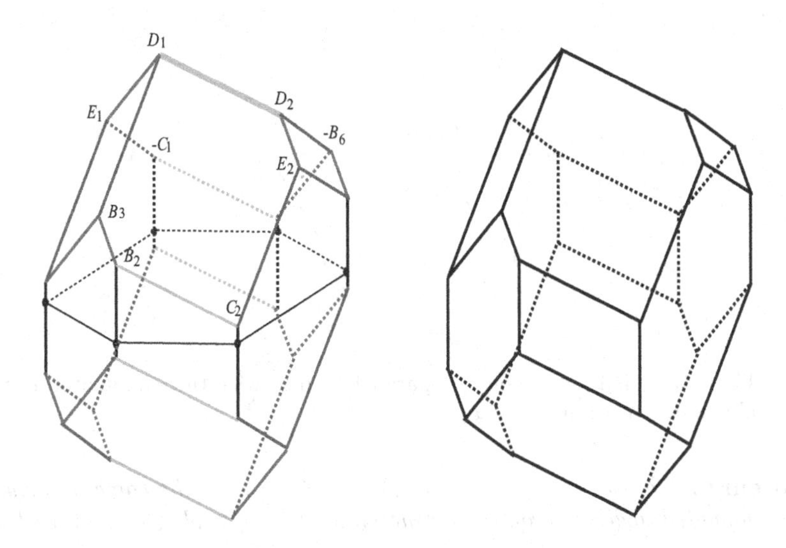

Fig. 4. Process to obtain a truncated octahedron

edges, we obtain the unique parallelohedron with its combinatorial type K of the truncated octahedron and the given parameters.

Remark. For each parallelohedron with its combinatorial type of the truncated octahedron, there are six belts. So, at most six different 5-tuples of parameters may correspond to a-equivalent parallelohedra in Theorem 4.

4 Non-primitive Parallelohedra

By applying the method used in the proof of Theorem 4, we get the following results.

Theorem 5. *The set of affine classes $\mathcal{A}(K)$ with the combinatorial type K of the rhombic dodecahedron are parameterized by a 3-tuple (α, β, h) where*

$$0 < \alpha, 0 < \beta \leq (\pi - \alpha)/2, 0 < h.$$

Proof. Since all faces of parallelohedra with the combinatorial type K of the rhombic dodecahedron are parallelograms, Step 3 in the proof of Theorem 4, we take a parallelogram $B_1B_2B_4B_5$ instead of the hexagon $B_1B_2 \cdots B_6$. Then we get a figure of the orthogonal projection to the xy-plane showed in Fig. 5. By drawing edges, we obtain the unique parallelohedron with combinatorial type K and the given parameters.

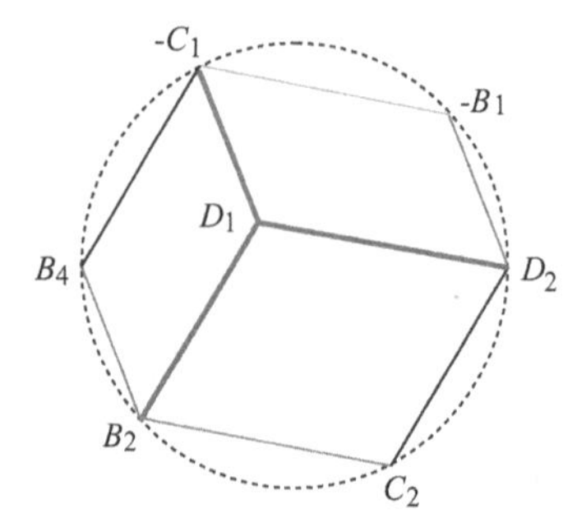

Fig. 5. The orthogonal projection of a parallelohedron with the combinatorial type of the rhombic dodecahedron to the xy-plane

Theorem 6. *The set of affine classes $\mathcal{A}(K)$ where K is the combinatorial type of the elongated dodecahedron is parameterized by a 4-tuple (α, β, h, l) where*

$$0 < \alpha,\ 0 < \beta \leq (\pi - \alpha)/2, 0 < h, 0 < l$$

.

Proof. It is proved by Theorem 6.

Acknowledgement. The authors would like to express their thanks to the referee for his careful reading and valuable suggestions. Especially he pointed out the lack of conditions in a preliminary version of Theorem 4.

References

1. Fedorov, E.S.: An introduction to the Theory of Figures, St. Petersburg (1885) (in Russian)
2. Minkowski, H.: Allgemeine Lerätze über die convexen Polyeder. Gött. Nachr., 198–219 (1897)
3. Venkov, B.A.: On a class of Euclidean Polyhedra. Vestn. Leningr. Univ., Ser. Mat. Fiz. 9, 11–31 (1954)
4. Voronoi, G.: Nouvelles applications des paramètres continus á la théorie des formes quadratiques Deuxiéme memoire: Recherches sur les parallélоédres primitifs. J. Reine Angew. Math. 134, 198–287 (1908); 136, 67–178 (1909)
5. Michel, L., Ryshkov, S.S., Senechal, M.: An extension of Voronoï's theorem on primitive parallelotopes. Europ. J. Combinatorics 16, 59–63 (1995)
6. Dolbilin, N., Itoh, J.-I., Nara, C.: Affine equivalent classes of parallelohedra. In: Akiyama, J., Bo, J., Kano, M., Tan, X. (eds.) CGGA 2010. LNCS, vol. 7033, pp. 55–60. Springer, Heidelberg (2011)

On Complexity of Flooding Games on Graphs with Interval Representations

Hiroyuki Fukui[1], Yota Otachi[1],
Ryuhei Uehara[1], Takeaki Uno[2], and Yushi Uno[3]

[1] Japan Advanced Institute of Science and Technology (JAIST),
Nomi, Ishikawa 923-1292, Japan
{s1010058,otachi,uehara}@jaist.ac.jp
[2] National Institute of Informatics (NII), Chiyoda-ku, Tokyo 101-8430, Japan
uno@nii.jp
[3] Osaka Prefecture University, Naka-ku, Sakai 599-8531, Japan
uno@mi.s.osakafu-u.ac.jp

Abstract. The flooding games, which are called Flood-It, Mad Virus, or HoneyBee, are a kind of coloring games and they have been becoming popular online. In these games, each player colors one specified cell in his/her turn, and all connected neighbor cells of the same color are also colored by the color. This flooding or coloring spreads on the same color cells. It is natural to consider the coloring games on general graphs: Once a vertex is colored, the flooding flows along edges in the graph. Recently, computational complexities of the variants of the flooding games on several graph classes have been studied. We investigate the one player flooding games on some graph classes characterized by interval representations. Our results state that the number of colors is a key parameter to determine the computational complexity of the flooding games.

1 Introduction

The *flooding game* is played on a precolored board, and each player colors a cell on the board in a turn. When a cell is colored with the same color as its neighbor, they will be merged into one colored area. If a player changes the color of one of the cells belonging to a colored area of the same color, the color of all cells in the area are changed. On the one player flooding game, it finishes when all cells are colored with one color. The objective of the game is to minimize the number of turns (or to finish the game within a given number of turns). This one player flooding game is known as Flood-It (Fig. 1). In Flood-It, each cell is a precolored square, the board consists of $n \times n$ cells, the player always changes the color of the top-left corner cell, and the goal is to minimize the number of turns. This game is also called Mad Virus played on a honeycomb board (Fig. 2). One can play both the games online (Flood-It (`http://floodit.appspot.com/`) and Mad Virus (`http://www.bubblebox.com/play/puzzle/539.htm`)).

In the original flooding games, the player colors a specified cell. However, it is natural to allow the player to color any cell. The original game is called *fixed* and

J. Akiyama, M. Kano, and T. Sakai (Eds.): TJJCCGG 2012, LNCS 8296, pp. 73–84, 2013.

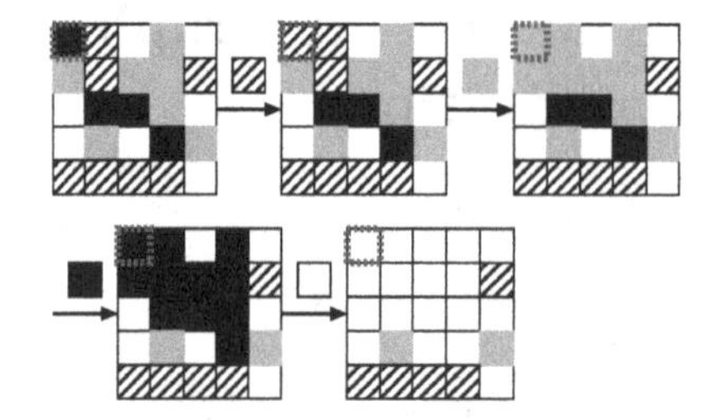

Fig. 1. A sequence of four moves on a 5×5 Flood-It board

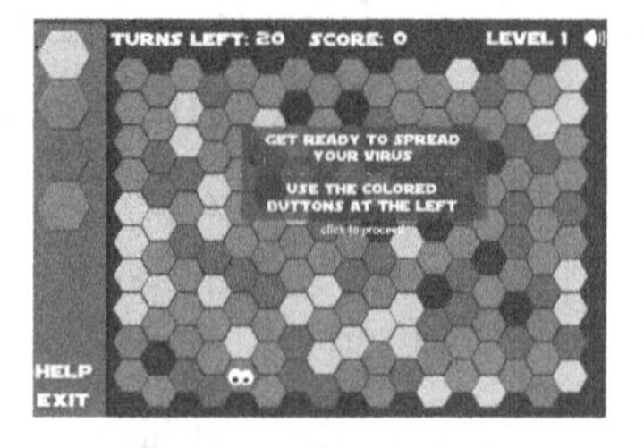

Fig. 2. The initial screen of the Mad Virus (`http://www.bubblebox.com/play/puzzle/539.htm`). The player changes the cell having eyes.

this extended game is called *free*. The flooding games are intractable in general on the grid board; both free and fixed versions are NP-hard on rectangular $3 \times n$ boards when the number of colors is 4 [MS12b], and the free version is still NP-hard on rectangular $2 \times n$ boards when the number of colors is $O(n)$ [MS11]. On the other hand, Meeks and Scott also show an $O(h(k)n^{18})$ time algorithm for the flooding game on $2 \times n$ boards when the number of colors is k, where $h(k)$ is an explicit function of k [MS11].[1]

In recent literature, the game board has been generalized to general graph; the vertex set corresponds to the set of precolored cells, and two cells are neighbors if and only if the corresponding vertices are adjacent in the graph. It is also natural to parameterize by the number k of colors. The generalized flooding games on general graphs have been well investigated from the viewpoint of computational complexity. We summarize recent results in Table 1. (The other related results can be found in [MS12a, CJMS12].)

Since the original game is played on a grid board, the extension to the graph classes having geometric representation is natural and reasonable. For example, each geometric object corresponding to a vertex can be regarded as a "power" or an "influence range" of the vertex. When a vertex is colored, the influence propagates according to the geometric representation. We first consider the case that the propagation is in one dimensional. That is, we first investigate the computational complexities of the flooding game on graphs that have interval representations. We will show that even in this restricted case, the problem is already intractable in general.

From the viewpoint of the geometric representation of graphs, the notion of interval graphs is a natural extension of paths. (We here note that path also models rectangular $1 \times n$ boards of the original game.) A path is an interval graph such that each vertex has an influence on at most two neighbors. In this case, the flooding games can be solved in polynomial time [LNT11, FNU$^+$11, MS12b]. However, we cannot extend the results for a path to an interval graph straightforwardly. There are two differences between paths and interval graphs:

[1] Recently, Clifford et al. give a linear time algorithm for $2 \times n$ boards with any number of colors [CJMS12].

Table 1. Computational complexities of the flooding games on some graph classes, where n and k are the number of vertices and colors, respectively

Graph classes	fixed	fixed, k is bounded
general graphs	NP-C	NP-C if $k \geq 3$ [ACJ+10] P if $k \leq 2$ (trivial)
(□/△/hex.) grids	NP-C	NP-C if $k \geq 3$ [LNT11]
paths/cycles	$O(n^2)$ [LNT11]	$O(n^2)$ [LNT11]
co-comparability graphs	P [FW12]	P [FW12]
split graphs	NP-C [FW12]	P [FW12]
caterpillars	P[2]	$O(4^k k^2 n^3)$ (This)
proper interval graphs	P[2]	$O(4^k k^2 n^3)$ (This)
interval graphs	P[2]	$O(4^k k^2 n^3)$ (This)
Graph classes	**free**	**free, k is bounded**
general graphs	NP-C	NP-C if $k \geq 3$ [ACJ+10] P if $k \leq 2$ [Lag10, LNT11]
(□/△/hex.) grids	NP-C	NP-C if $k \geq 3$ [LNT11]
paths/cycles	$O(n^3)$ [FNU+11][3]	$O(n^3)$ [FNU+11][3]
split graphs	NP-C (This)	$O((k!)^2 + n)$ (This)
caterpillars	NP-C (This,[MS11])	$O(4^k k^2 n^3)$ (This)
proper interval graphs	NP-C (This)	$O(4^k k^2 n^3)$ (This)
interval graphs	NP-C (This)	$O(4^k k^2 n^3)$ (This)

Interval graphs have branches and twins. Interestingly, one of them is sufficient to make the flooding game intractable:

Theorem 1. *The free flooding game is* NP*-complete even on proper interval graphs, or even on caterpillars. These results still hold even if the maximum degree of the graphs is bounded by 3.*

Both of the classes consist of very simple interval graphs. The results are tight since they degenerate to the set of paths when the maximum degree is bounded by 2. On the other hand, when the number of colors is a constant k, the game on interval graphs becomes tractable.

Theorem 2. *The free flooding game on an interval graph can be solved in* $O(4^k k^2 n^3)$ *time.*

Thus the game is fixed parameter tractable with respect to the number of colors.

We also extend the results for the fixed flooding game on a split graph mentioned in [FW12] to the free flooding game on a split graph. Precisely, the free flooding game is NP-complete even on a split graph, and it can be solved in $O((k!)^2 + n)$ time when the number k of colors is fixed.

[2] The class of co-comparability graphs properly contains interval graphs and hence caterpillars and proper interval graphs. Since this game is polynomial time solvable on a co-comparability graph, so they follow.

[3] In [FNU+11], the authors gave an $O(kn^3)$ algorithm. However, it can be improved to $O(n^3)$ easily in the same way in [LNT11].

Although we only consider one player game in this paper, it is also natural to extend to multi-players. Two-player variant is known as HoneyBee, which is available online at `http://www.ursulinen.asn-graz.ac.at/Bugs/htm/games/biene.htm`. Fleischer and Woeginger have investigated this game from the viewpoint of computational complexity. See [FW12] for further details.

Due to the space limitation, we omit the details of some proofs, which can be found in [FOU+13].

2 Preliminaries

We model the flooding game in the following graph-theoretic manner. The game board is a connected, simple, loopless, undirected graph $G = (V, E)$. We denote by n and m the number of vertices and edges, respectively. There is a set $C = \{1, 2, \ldots, k\}$ of colors, and every vertex $v \in V$ is precolored (as input) with some color $col(v) \in C$. Note that we may have an edge $\{u, v\} \in E$ with $col(u) = col(v)$. For a vertex set $U \subseteq V$, the vertex induced graph $G[U]$ is the graph (U, F) with $F = E \cap \{\{u, v\} \mid u, v \in U\}$. For a color $c \in C$, the subset V_c contains all vertices in V of color c. For a vertex $v \in V$ and color $c \in C$, we define the *color-c-neighborhood* $N_c(v)$ by the set of vertices in the same connected component as v in $G[V_c]$. Similarly, we denote by $N_c(W) = \cup_{w \in W} N_c(w)$ the color-c-neighborhood of a subset $W \subseteq V$. For a given graph $G = (V, E)$ and the precoloring $col()$, a *coloring operation* (v, c) for $v \in V$ and $c \in C$ is defined by, for each vertex $v' \in N_{c'}(v)$ with $c' = col(v)$, setting $col(v') = c$. For a given graph $G = (V, E)$ and a sequence $(v_1, c_1), (v_2, c_2), \ldots, (v_t, c_t)$ of coloring operations in $V \times C$, we let $G_0 = G$ and G_i is the graph obtained by the coloring operation (v_i, c_i) on G_{i-1} for each $i = 1, 2, \ldots, t$. In the case, we denote by $G_{i-1} \rightarrow_{(v_i, c_i)} G_i$ and $G_0 \rightarrow^i G_i$ for each $0 \leq i \leq t$. Then the problem in this paper are defined as follows[4]:

Problem 1. Free flooding game

Input : A graph $G = (V, E)$ such that each vertex in V is precolored with $col(v) \in C$ and an integer t;

Output: Determine if there is a sequence of coloring operations $((v_1, c_1), (v_2, c_2), \ldots, (v_t, c_t))$ of length t such that all vertices in the resulting graph G' (i.e. $G \rightarrow^t G'$) have the same color;

For the problem, if a sequence of operations of length t colors the graph, the sequence is called a *solution* of length t.

A graph (V, E) with $V = \{v_1, v_2, \cdots, v_n\}$ is an *interval graph* if there is a set of (real) intervals $\mathcal{I} = \{I_{v_1}, I_{v_2}, \cdots, I_{v_n}\}$ such that $\{v_i, v_j\} \in E$ if and only if $I_{v_i} \cap I_{v_j} \neq \emptyset$ for each i and j with $1 \leq i, j \leq n$ (Fig. 3(a)(b)). We call the set $\mathcal{I}$ of intervals an *interval representation* of the graph. For each interval I, we denote by $L(I)$ and $R(I)$ the left and right endpoints of the interval, respectively

[4] In the fixed flooding game, $v_1 = v_2 = \cdots = v_t$ is also required, and this vertex is specified as a part of input.

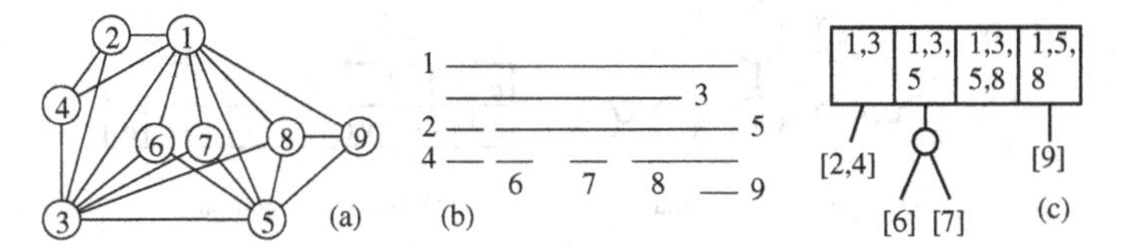

Fig. 3. (a) An interval graph G, (b) one of interval representations of G, and (c) unique $\mathcal{MPQ}$-tree of G (up to isomorphism)

(hence we have $L(I) \leq R(I)$ and $I = [L(I), R(I)]$). For a point p, let $N[p]$ denote the set of intervals containing the point p. In general, there exist many interval representations for an interval graph G. On the other hand, there exists unique representation for an interval graph G, which is called $\mathcal{MPQ}$-*tree* of G. The definition of $\mathcal{MPQ}$-tree is postponed to Section 3.2.

An interval representation is *proper* if no two distinct intervals I and J exist such that I properly contains J or vice versa. An interval graph is *proper* if it has a proper interval representation. If an interval graph G has an interval representation $\mathcal{I}$ such that every interval in $\mathcal{I}$ has the same length, G is said to be a *unit* interval graph. Such an interval representation is called a *unit* interval representation. It is well known that the class of proper interval graphs coincides with the class of unit interval graphs [Rob69]. That is, given a proper interval representation, we can transform it into a unit interval representation. A simple constructive way of the transformation can be found in [BW99]. With perturbations if necessary, we can assume without loss of generality that $L(I) \neq L(J)$ (and hence $R(I) \neq R(J)$), and $R(I) \neq L(J)$ for any two distinct intervals I and J in a unit interval representation $\mathcal{I}$.

A connected graph $G = (V, E)$ is a *caterpillar* if V can be partitioned into B and H such that $G[B]$ is a path, H is an independent set, and each vertex in H is incident to exactly one vertex in B. It is easy to see that the caterpillar G is an interval graph. We call B (and $G[B]$) *backbone*, and each vertex in H *hair* of G, respectively. A graph $G = (V, E)$ is a *split graph* if V can be partitioned into C and I such that C is a clique and I is an independent set. (A vertex set C is *clique* if every pair of vertices is joined by an edge, and it is *independent set* if no pair is joined.)

3 Graphs with Interval Representations

Let $G = (V, E)$ be an interval graph precolored with at most k colors. We first show that, when k is not bounded, the flooding game on G is NP-complete even if G is a caterpillar or a proper interval graph. Next we show an algorithm that solves the flooding game in $O(4^k k^2 n^3)$ on a proper interval graph. Lastly, we extend the algorithm to general interval graphs. That is, the flooding game is fixed parameter tractable on an interval graph with respect to the number of colors.

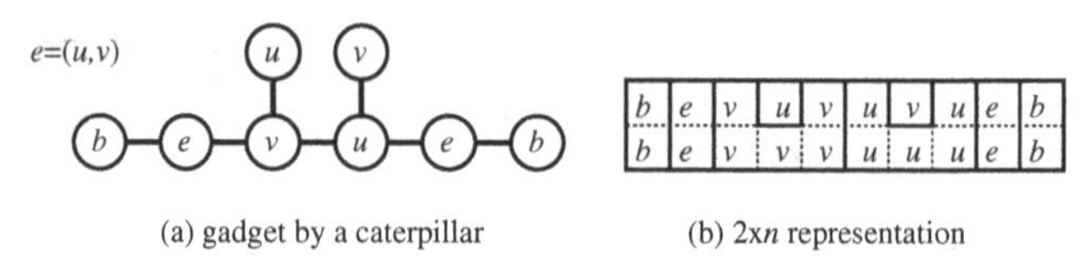

(a) gadget by a caterpillar (b) 2xn representation

Fig. 4. A gadget for $e = (u, v)$

3.1 NP-Completeness on Simple Interval Graphs

To prove Theorem 1, we reduce the following well-known NP-complete problem to our problems (see [GJ79, GT1]):

Problem 2. Vertex Cover

Input : A graph $G = (V, E)$, and an integer k;
Output: Determine if there is a subset S of V such that for each edge $e = \{u, v\} \in E$, $e \cap S \neq \emptyset$ and $|S| = k$;

Let $G = (V, E)$ and k be an instance of the vertex cover problem. Let $n = |V|$ and $m = |E|$.

Caterpillar. We first construct a caterpillar[5]. The key gadget is shown in Fig. 4(a). We replace an edge $e = (u, v)$ by a path $(b_1, b_2, b_3, b_4, b_5, b_6)$ with two hairs h_3 and h_4 attached to b_3 and b_4. The colors are as shown in the figure: $col(b_1) = col(b_6) = b$, $col(b_2) = col(b_5) = e$, $col(b_3) = col(h_4) = v$, and $col(b_4) = col(h_3) = u$. This gadget cannot be colored in at most three turns. On the other hand, there are some ways to color them in four turns. One of them is: color b_3 by u, color b_3 by e, color b_3 by b, and color h_4 by b.

Now we turn to the reduction from a general graph (Fig. 5). We first arrange the edges in arbitrary way, and replace each edge by the gadget in Fig. 4(a). In this time, each vertex of color b is shared by two consecutive edges. In other words, endpoints of the gadget are shared by two consecutive gadget except both ends. This is the reduction. The resulting graph is a caterpillar, the reduction is a polynomial-time reduction, and the flooding game is clearly in NP. Thus it is sufficient to show that a minimum vertex cover S of G gives a solution with $3m + |S|$ operatoins of the flooding game on the resulting graph and vice versa.

As shown in the example, all vertices on the backbone are colored by b in $3m$ coloring operations. On the other hand, $3m$ coloring operations are required to color the backbone since each gadget is separated by the color b. Moreover, we have a leftover hair at each gadget, and their colors form a vertex cover S since they hit all edges. Therefore, once we have a vertex cover S, we color the resulting graph with $3m + |S|$ operations. On the other hand, if we can color the resulting graph with $3m + |S'|$ operations, we can extract $3m$ operations to color the backbone, and each of $|S'|$ operations is an operation to color a leftover hair,

[5] We sometimes identify an interval graph and its interval representation.

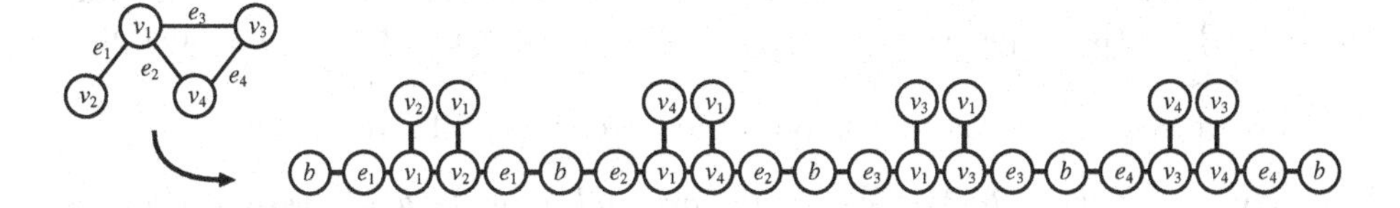

Fig. 5. An example of reduction to a caterpillar

which gives us a vertex cover. Therefore, the graph G has a vertex cover of size k' iff the resulting graph can be colored with $3m + k'$ coloring operations.

On the other hand, if we have a vertex cover S of the original graph, we can color the resulting graph in $3m + |S|$ operations by joining the backbones with leaving hairs corresponding to S.

We note that the basic idea of this reduction can be found in the proof of the NP-completeness on rectangular $2 \times n$ boards in [MS11]. In fact, the gadget in Fig. 4(a) can be represented by a rectangular $2 \times n$ board shown in Fig. 4(b), and we can obtain essentially the same proof in [MS11]. We here explained the details of the proof to make this paper self-contained, and this idea is extended to proper interval graphs in the next section.

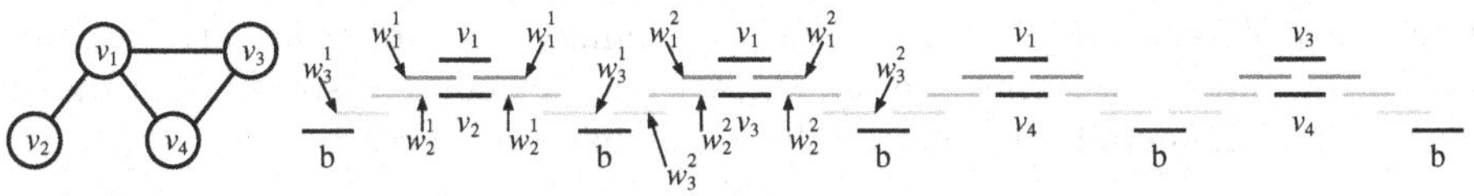

Fig. 6. Reduction from Vertex Cover to Flooding game

Proper Interval Graph. From a given general graph, we next construct an interval representation $\mathcal{I}$ of a proper interval graph as follows (Fig. 6).
(1) C is the color set $V \cup \{w_i^j \mid 1 \le i \le m-1, 1 \le j \le m\} \cup \{b\}$ of $n+m(m-1)+1$ different colors. (Note that each vertex in V has its own unique color.)
(2) For each $0 \le i \le m$, we put an interval $I_i = [4i, 4i+1]$ with precolor $col(I) = b$. We call these $m+1$ intervals *backbones*.
(3) For each $e_i = \{u, v\} \in E$ with $0 \le i < m$, we add two identical intervals $J_i = [4i+2, 4i+3]$ and $J_i' = [4i+2, 4i+3]$ with precolor $col(J_i) = u$ and $col(J_i') = v$. (Note that the ordering of the edges is arbitrary.)
(4) Each two identical intervals J_i and J_i' are connected to the left and right backbone by paths of length m. Precisely, a left backbone $I = [4i, 4i+1]$ and the two intervals $J_i = [4i+2, 4i+3]$ and $J_i' = [4i+2, 4i+3]$ are joined by a path $(w_m^i, w_{m-1}^i, \ldots, w_1^i, w_0^i)$, where $I = [4i, 4i+1] = I_{w_m^i}$ and $J_i = I_{w_0^i}$ (which is identical to J_i'). (Note that w_1^i has three neighbors: w_2^i and two vertices corresponding to J_i and J_i'.) The intervals $J_i = [4i+2, 4i+3]$, $J_i' = [4i+2, 4i+3]$ are connected to the right backbone $I = [4i+4, 4i+5]$ in a symmetric way. That is, they are connected by a path $(w_0^i, w_1^i, \ldots, w_{m-1}^i, w_m^i)$ such that $I_{w_0^i} = [4i+2, 4i+3]$, $I_{w_m^i} = [4i+4, 4i+5]$. For each j with $1 \le j \le m-1$, we set $col(w_j^i) = w_j^i$ with $1 \le j \le n$. That is, two paths from $[4i+2, 4i+3]$ to

both backbones have the same color sequence, and when we color the interval J_i (or J_i') by the sequence $w_1^i, \ldots, w_{m-1}^i, b$, we can connect the left and right backbones.

Now we show a lemma that implies the latter half of Theorem 1.

Lemma 1. *In the reduction above, the original graph G has a vertex cover of size k' if and only if there is a sequence of coloring operations of length $m^2 + k'$ to make the resulting interval representation in monochrome.*

Proof. (Outline.) We first suppose that the graph G has a vertex cover S of size k'. Then we can construct a sequence of coloring operations of length $m^2 + k'$ as follows. First step is joining the backbones. Let $e_i = \{u, v\}$ be an edge in E. Since S is a vertex cover, without loss of generality, we assume $u \in S$. Then we color v by $w_1^i, w_2^i, \ldots, w_{m-1}^i$, and b (we do not mind if v is in S or not). Repeat this process for every edge. Then all the backbones are connected and colored by b after m^2 colorings. We then still have m intervals corresponding to the vertices in S. Thus we pick up each vertex v in S and color the backbone by $col(v)$. After $|S|$ colorings, all vertices become monochrome.

Next we suppose that we have a sequence of coloring operations of length $m^2 + k'$ that makes the representation monochrome. Then, among them, we have to use m^2 operations to join the backbones. Moreover, they should start from each of twin intervals. After that, remaining ones of twins give us the vertex cover of size k'. Thus, we can extract a vertex cover of size k' from these operations. The details are omitted due to the lack of space. □

3.2 Polynomial Time Algorithm on Interval Graphs for Fixed Number of Colors

We first show an algorithm for proper interval graphs that runs in polynomial time if the number of colors is fixed. Next we extend the algorithm to deal with general interval graphs.

Algorithm for a Proper Interval Graph. Let $\mathcal{I}(G)$ be an interval representation of the proper interval graph $G = (V, E)$. The interval representation is given in a compact form (see [UU07] for details). Precisely, each endpoint is a positive integer, $N[p] \neq N[p+1]$ for each integer p, and there are no indices $N[p] \subset N[p+1]$ or vice versa for each integer p with $N[p] \neq \emptyset$ (otherwise we can shrink it). Intuitively, each integer point corresponds to a set of different endpoints of the intervals since the representation has no redundancy. Then, it is known that $\mathcal{I}(G)$ is unique up to isomorphism when G is a proper interval graph (see [SYKU10]), and $\mathcal{I}(G)$ can be placed in $[0..P]$ for some $P \leq 2n - 1$. Sweeping a point p from 0 to P on the representation, the color set $N[p]$ differs according to p. More precisely, we obtain $2P + 1$ different color sets for each $p = 0, 0.5, 1, 1.5, 2, 2.5, \ldots, P - 0.5, P$. We note that for each integer p, $N[p+0.5] \subset N[p]$ and $N[p+0.5] \subset N[p+1]$. Let S_i be the color set obtained by the ith p (to simplify the notation, we use from S_0 to S_{2P}). Since the color set

C has size k, each S_i consists of at most k colors. That is, the possible number of color sets is $2^k - 1$ (since $S_i \neq \emptyset$).

Now we regard the unique interval representation as a path $\mathcal{P} = (\hat{S_0}, \hat{S_1}, \ldots, \hat{S_{2P}})$, where each vertex $\hat{S_i}$ is precolored by the color set S_i. Then we can use a dynamic programming technique, which is based on the similar idea to the algorithms for the flooding game on a path in [FNU+11, LNT11, MS12b]. On a path, the correctness of the strategy comes from the fact that removing the color at the point p divides the interval representation into left and right, and they are independent after removing the color at the point p. However, on $\mathcal{P}$, we have to take care of the influence of changing the color set of a vertex in the original interval graph. In the algorithms for an ordinary path, changing the color of a vertex has an influence to just two neighbors. In our case, when we change a color c in S_i at a point p to c', all reachable color sets joined by c from $\hat{S_i}$ are changed. Thus we have to remove c from S_j and add c' to S_j for each j with $i' \leq j \leq i''$, where i' and i'' are the leftmost and the rightmost vertices such that $c \in \cap_{i' \leq j \leq i''} S_j$. By this coloring operation, some colors may be left independent on the backbone of color c'. To deal with these color sets, we maintain a table $f(\ell, r, c, S)$ that is the minimum number of coloring operations to satisfy the following conditions: (1) $c \in S_i$ for each i with $\ell \leq i \leq r$, and (2) $\cup_{\ell \leq i \leq r} S_i \subseteq (S \cup \{c\})$. That is, $f(\ell, r, c, S)$ gives the minimum number of coloring operations to make this interval connected by the color c, and the remaining colors in this interval are contained in S. Once we obtain $f(0, 2P, c, S)$ for all c and S on $\mathcal{P}$, we can obtain the solution by the following lemma:

Lemma 2. *For a given proper interval graph G, let $f(\ell, r, c, S)$ be the table defined above. Then the minimum number of coloring operations to make G monochrome is given by* $\min_c f(0, 2P, c, \phi) = \min_{c,S}(f(0, 2P, c, S) + |S|)$.

Proof. (Outline) We first show that we can make G monochrome within $\min_{c,S}(f(0, 2P, c, S) + |S|)$ coloring operations. For each c and S, by the definition of the table, we can make that every color set S_i contains c with $f(0, 2P, c, S)$ coloring operations. This means that every vertex v is either $col(v) = c$ or $N(v)$ contains some u such that $col(u) = c$, and $G[N_c(v)]$ is connected. Therefore, taking each color $c' \in S$, and changing the color of any vertex of color c to c', all vertices of color c' and c are merged to the vertices of color c'. Therefore, repeating this process, we can make G monochrome with $\min_{c,S}(f(0, 2P, c, S) + |S|)$ coloring operations. It is easy to observe that we have $\min_c f(0, 2P, c, \phi) = \min_{c,S}(f(0, 2P, c, S) + |S|)$.

We next show that the above strategy cannot be improved. In the definition of the function f, we take a strategy that (1) first, color the interval $[i, j]$ with a color c and (2) second, color the remaining colors in the interval $[i, j]$ by changing the color c. We say that this color c *dominates* the interval $[i, j]$ after the first step. We suppose that a coloring operation pick up a color c of a vertex v and change it to another color c'. Then the color sets in an interval $[i, j]$ are changed since $i \leq L(I_v) \leq R(I_v) \leq j$, i is the leftmost vertex such that S_i contains c, and j is the rightmost vertex with $c \in S_j$. Here we suppose that $[i, j]$ is properly contained in another interval $[i', j']$ with $i' \leq i \leq j \leq j'$ that is dominated by a color c''. Then, we can override to use the color c'' instead of c in the sense that changing c to c' is not better than changing c'' to c'. That is, when we change a color c to c', if there is another overriding color c'', it is not worse to change a color c'' to c' instead of c. Repeating this argument, we can see that the above strategy is not worse any other strategy. Thus we cannot improve it. □

By careful case analysis, this function satisfies the following recursive relation.

$$
\begin{array}{ll}
f(\ell, r, c, S) = \min\{ & \\
\min_{\ell < i \le r, c' \in C \setminus \{c\}} f(\ell, i-1, c, S') + f(i, r, c', S'') + 1 & \text{such that } S', S'' \subseteq S \cup \{c\} \\
\min_{\ell < i \le r,, c' \in C \setminus \{c\}} f(\ell, i-1, c', S') + 1 + f(i, r, c, S'') & \text{such that } S', S'' \subseteq S \cup \{c\} \\
\min_{\ell < i \le r} f(\ell, i-1, c, S') + f(i, r, c, S'') & \text{such that } S', S'' \subseteq S \\
\} &
\end{array}
$$

The correctness of the dynamic programming algorithm based on this recursive function is given by Lemma 2. Thus the remaining task is showing the computational complexity of the function. This can be done in a standard dynamic programming technique, but the details are omitted due to lack of space.

Lemma 3. *The value of* $\min_c f(0, 2P, c, \phi)$ *can be computed in* $O(4^k k^2 n^3)$ *time.*

We here note that a similar concept can be found in the algorithm for $2 \times n$ board in [MS11].

Extension to Interval Graphs. A proper interval graph has a simple interval representation. Especially, its interval representation is linear and essentially unique up to isomorphism. Therefore we can use the dynamic programming technique on the unique path-like structure. On the other hand, a general interval graph has exponentially many different interval representations. To deal with an interval graph, we use a tree representation that was used to solve the graph isomorphism problem for interval graphs [KM89]. The *$\mathcal{MPQ}$-tree* stands for *modified $\mathcal{PQ}$-tree*, and this notion was introduced by Korte and Möhring in [KM89]. For an interval graph, the $\mathcal{MPQ}$-tree is uniquely determined up to isomorphism. To solve the flooding problem on an interval graph, we extend the algorithm for proper interval graph to solve the problem on the $\mathcal{MPQ}$-tree of color sets. The $\mathcal{MPQ}$-tree maintains inclusion relationships among intervals. That is, if an interval I appears in a node p that is an ancestor of another node q, all intervals appearing in the node q is properly contained in the interval I. Thus, by the same argument of the proof of Lemma 2, a coloring operation for the intervals appearing in the node q can be overridden by the coloring operation of I. Therefore, in the same manner of the algorithm for the color sets of proper interval graphs, it is enough to consider the coloring operation of I, and the other intervals properly contained in I will be dealt with as the set of remaining colors in S in the function $f(0, 2P, c, S)$ in the previous algorithm. However, we omit the details in this draft due to lack of space.

4 Split Graphs

In [FW12], the fixed flooding game on a split graph is investigated. Using a similar idea in [FNU+11], we can extend the results for the fixed flooding game to the free flooding game.

Theorem 3. *(1) The free flooding game is* NP*-complete even on a split graph. (2) The free flooding game on a split graph can be solved in* $O((k!)^2 + n)$ *time, where* n *and* k *are the number of vertices and colors, respectively.*

Proof. (Sketch) (1) In [FW12], the feedback vertex set problem is reduced to the fixed flooding game on a split graph $G = (V, E)$. The resulting graph G consists of a clique K and an independent set I. Each vertex in I has degree one except one universal vertex u incident to all vertices in K. It is easy to see that this universal vertex u can be one of the clique K. Now we add $|K|$ vertices to I and join them to u, and each of them is colored by $|K|$ colors that are same to the colors of vertices in K. Then, the resulting graph is still split graph with clique $K \cup \{u\}$. We consider the free flooding game on this new split graph. Then, using the similar argument in [FNU+11], this graph has a solution of a given length if and only if there is a sequence of operations that always colors the universal vertex u. Thus the feedback vertex set problem has a solution if and only if the free flooding game has a solution.

(2) We can observe that there is a solution of length at most $2k$ that first makes all vertices in K having the same color, and changes the color of the clique to join the vertices in I. We can also see that there is an optimal solution of this form. This means that we always change the color of a clique vertex. Since the vertices in K of the same color are always connected, the number of possibilities of each operation is at most $k'(k'-1)$, where k' is the current number of colors used in K. Thus, we can find an optimal solution in $O((k!)^2 + n)$ time. □

Acknowledgment. The authors thank Eric Theirry for sending [LNT11]. The authors also thank to the anonymous referee, who gave fruitful suggestions for improve the presentation of this paper.

References

[ACJ+10] Arthur, D., Clifford, R., Jalsenius, M., Montanaro, A., Sach, B.: The Complexity of Flood Filling Games. In: Boldi, P. (ed.) FUN 2010. LNCS, vol. 6099, pp. 307–318. Springer, Heidelberg (2010)

[BW99] Bogart, K.P., West, D.B.: A short proof that 'proper=unit'. Discrete Mathematics 201, 21–23 (1999)

[CJMS12] Clifford, R., Jalsenius, M., Montanaro, A., Sach, B.: The Complexity of Flood Filling Games. Theory of Computing Systems 50, 72–92 (2012)

[FNU+11] Fukui, H., Nakanishi, A., Uehara, R., Uno, T., Uno, Y.: The Complexity of Free Flood Filling Game. In: WAAC 2011, pp. 51–56 (2011)

[FOU+13] Fukui, H., Otachi, Y., Uehara, R., Uno, T., Uno, Y.: On Complexity of Flooding Games on Graphs with Interval Representations. arXiv:1206.6201 (January 28, 2013)

[FW12] Fleischer, R., Woeginger, G.J.: An algorithmic analysis of the Honey-Bee game. Theoretical Computer Science 452, 75–87 (2012)

[GJ79] Garey, M.R., Johnson, D.S.: Computers and Intractability — A Guide to the Theory of NP-Completeness. Freeman (1979)

[KM89] Korte, N., Möhring, R.H.: An Incremental Linear-Time Algorithm for Recognizing Interval Graphs. SIAM Journal on Computing 18(1), 68–81 (1989)

[Lag10] Lagoutte, A.: 2-Free-Flood-It is polynomial. Technical report, arXiv:1008.3091v1 (2010)

[LNT11] Lagoutte, A., Naual, M., Thierry, E.: Flooding games on graphs. In: Latin-American Algorithms, Graphs and Optimization Symposium (LAGOS 2011) (2011)

[MS11] Meeks, K., Scott, A.: The complexity of Free-Flood-It on $2 \times n$ boards. arXiv:1101.5518v1 (January 2011)

[MS12a] Meeks, K., Scott, A.: Spanning Trees and the Complexity of Flood-Filling Games. In: Kranakis, E., Krizanc, D., Luccio, F. (eds.) FUN 2012. LNCS, vol. 7288, pp. 282–292. Springer, Heidelberg (2012)

[MS12b] Meeks, K., Scott, A.: The complexity of flood-filling games on graphs. Discrete Applied Mathematics 160(7-8), 959–969 (2012)

[Rob69] Roberts, F.S.: Indifference graphs. In: Harary, F. (ed.) Proof Techniques in Graph Theory, pp. 139–146. Academic Press (1969)

[SYKU10] Saitoh, T., Yamanaka, K., Kiyomi, M., Uehara, R.: Random Generation and Enumeration of Proper Interval Graphs. IEICE Transactions on Information and Systems E93-D(7), 1816–1823 (2010)

[UU07] Uehara, R., Uno, Y.: On Computing Longest Paths in Small Graph Classes. International Journal of Foundations of Computer Science 18(5), 911–930 (2007)

How to Generalize Janken – Rock-Paper-Scissors-King-Flea

Hiro Ito

School of Informatics and Engineering,
The University of Electro-Communications,
Chofu, Tokyo 182-8585, Japan
itohiro@uec.ac.jp

Abstract. Janken, which is a very simple game and it is usually used as a coin-toss in Japan, originated in China, and many variants are seen throughout the world. A variant of janken can be represented by a tournament, where a vertex corresponds a sign and an arc (x, y) means sign x defeats sign y. However, not all tournaments define useful janken variants, i.e., some janken variants may include a useless sign, which is strictly inferior than another sign in any case. We first shows that for any positive integer n except 2 and 4, we can construct a janken variant with n signs without useless signs. Next we introduces a measure of amusement of janken variants by using the variation of the difference of out-degree and in-degree. Under this measure, we show that a janken variant has the best amusement among ones with n signs if and only if it corresponds to one of the tournaments defined by J. W. Moon in 1993. Following these results, we present a janken variant "King-fles-janken," which is the best amusing janken variant among ones with five signs.

1 Introduction

Janken is a simple game to decide a winner by simultaneously holding out one hand in one of three gestures (signs) to signify *rock* (closed hand), *paper* (open hand), or *scissors* (closed hand with index and middle fingers extended). So, it's also called *rock-paper-scissors*. Rock defeats scissors, scissors defeats paper, and paper defeats rock. These relation can be represented by using a *tournament*, i.e., an asymmetric complete digraph, where an arc (x, y) means x defeats y as Fig. 1 (a).

Janken originated in China and many variants are seen throughout the world. For example, in the local rule in a part of France, *pot*[1] (forming a hole) is added to the signs, and hence they have four signs. Pot defeats rock and scissors (since they are sunk) but is defeated by paper (since it covers the mouse). This variant of janken can be also represented by a corresponding tournament, which has four vertices (see Fig. 1 (b)).

[1] Sometimes "well" is used in replace with pot.

J. Akiyama, M. Kano, and T. Sakai (Eds.): TJJCCGG 2012, LNCS 8296, pp. 85–94, 2013.

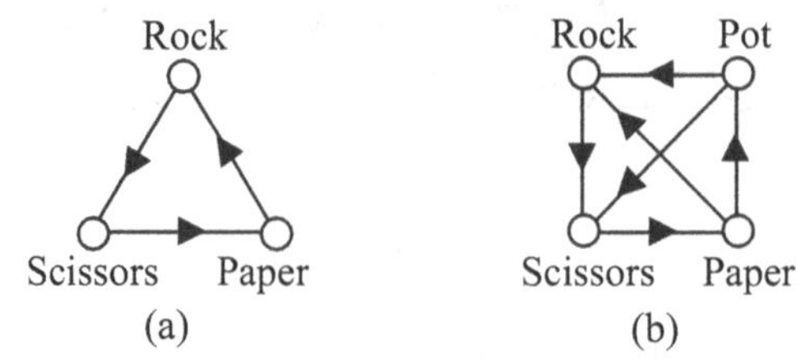

Fig. 1. Jankens represented by tournaments: (a) the popular janken, (b) pot-janken

A variant of janken can be defined corresponding to any tournament[2]. (For properties on tournaments, see [3].) However, not all tournaments define useful variants of janken. For example, the above defined extended janken (here we call it *pot-janken*) has a weak point, i.e., forming pot is always a more excellent selection than forming rock, since both defeats scissors, are defeated by paper, and pot defeats rock. Thus no one will use rock, and so pot-janken is essentially the same as the popular one.

Our Contribution. We call a variant of janken (or a tournament) is *efficient* if it contains no such useless sign, which will be rigorously defined in Definition 1 in Section 2. This idea is the same as the known idea "totally incomparable," i.e., a tournament is *totally incomparable* if and only if the corresponding janken is efficient.

In this paper, we first show that:

Theorem 1. *Let n be a positive integer. There is an efficient tournament (a janken variant) with n vertices (signs) if and only if $n \neq 2, 4$.*

Next, the level of "amusement" of janken variants is examined. We define the level of amusement of a janken variant as the irregularity of the corresponding tournament T. The irregularity of T is $\mathrm{irr}(T) := \sum_{x \in V}((\text{out-degree of } x) - (\text{in-degree of } x))^2$.

By using this idea, we obtain the following result:

Theorem 2. *Let T be an efficient tournament (janken variant) with $n \neq 2, 4$ vertices (signs). Then*

$$\mathrm{irr}(T) \leq \begin{cases} \dfrac{(n-1)(n-2)(n-3)}{3} & \text{if } n \text{ is odd,} \\[2ex] \dfrac{n^3 - 6n^2 + 11n - 48}{3} & \text{if } n \text{ is even,} \end{cases}$$

with equality holding if and only if T is one of the tournaments M_n, which are defined by J.W. Moon in 1993 [4].

[2] We don't consider any janken variant allowing ties between distinct signs in this paper.

Based on this result, we introduce a new janken variant called "King-flea-janken" which is the unique efficient janken variant having the highest irregularity (i.e., amusement).

Related Work. Various properties on tournaments are shown in [1], [3], and [4]. Variants of Janken were examined by Fisher and Ryan [2] as "tournament games," where optimal strategies are discussed. Note that "sign x is not useless" and "x has nonzero probability in any optimal mixed strategy" is not equivalent. Although in fact the farmer leads the latter, the reverse doesn't hold generally.

2 Janken Variants and Efficiency

2.1 Janken Variants and Terminology

Janken have many variants in history and in the world [5]: many of them consist of three signs, e.g., "rock-paper-scissors" are replaced with "frog-snake-slug" in Japan, "fox-rifle-headman" in Japan (it is called *kitsune-ken*, where kitsune means a fox), "tiger-Watounai (a hero)-his mother" in Japan (it is called *tora-ken*, where tora means a tiger), "tiger-soldier-commander" in Myanmar, or "ant-human-elephant" in India, etc. Some janken variants require gestures by the whole body, e.g., kitsune-ken and tora-ken are such type. These janken variants are all represented by the tournament of Fig. 1 (a), i.e., essentially the same, only except for gestures.

However, there are some janken variants consisting of more than three signs: As well as the above described pot-janken, in Malaysia they have a janken variant consisting of five signs, "rock-pistol-water-bird-board," and the relation between them is represented in Fig. 2 (a). In the web (e.g. [7]), we find "rock-paper-scissors-Spock-lizard," shown in Fig. 2 (b). Although these two janken variants are very similar, they are different: if the relation between pistol and bird is reversed, they become essentially the same.

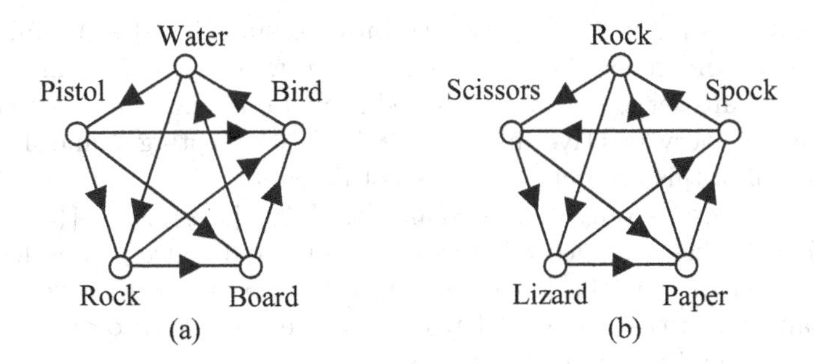

Fig. 2. Janken variants with five signs

The janken variant of Fig. 2 (a) also includes useless signs, bird and rock, as rock in the pot-janken investigated in Section 1. Here we define this word rigorously.

From here, a tournament is identified with the corresponding janken variant. A vertex of a tournament is also identified with the corresponding sign. A tournament consists of n vertices is sometimes denoted by an *n-tournament.*

Definition 1. *For a tournament $T = (V, A)$, if a pair of vertices x and y satisfies the following two conditions, then x is* superior *than y and y is* useless*:*

- *$(x, y) \in A$, and*
- *for any vertex $z \in V$, if $(y, z) \in A$, then $(x, z) \in A$,*

If a tournament includes no useless vertex, then it is said to be efficient *or* totally incomparable.

For example, in the pot-janken, rock is useless, since pot is superior than it, and hence we observe that the pot-janken is not efficient. Similarly in the janken variant in Fig. 2 (a), board is superior than bird and pistol is superior than rock. Thus this janken becomes simpler such as "pistol-water-board," which is essentially the same as "rock-paper-scissers."

For a vertex $x \in V$ of a tournament $T = (V, A)$, the number of edges outgoing from (incoming to) v is called *out-degree* (*in-degree*) of x and expressed by $\deg^+(x)$ ($\deg^-(x)$), i.e.,

$$\deg^+(x) := |\{y \in V \mid (x, y) \in A\}| \text{ and}$$
$$\deg^-(x) := |\{y \in V \mid (y, x) \in A\}|.$$

Note that $\deg^+(x)$ ($\deg^-(x)$) means the number of vertices which are defeated by x (which defeat x).

2.2 Existence of Efficient Jankens with n Signs

Here the first question is coming: can we construct an efficient tournament with n=4 or 5? For more generally, for what integer n we can do? For this question, we obtained an answer as Theorem 1 stated in Section 1, i.e., there is an efficient tournament for every positive integer $n \geq 1$, only excepting 2 and 4. We will show the proof of this theorem, which is not difficult.

For $n = 2$, there is only one tournament: $(V = \{x, y\}, A = \{(x, y)\})$, and clearly it is not efficient, since x is superior than y and hence y is useless.

The diameter of T is the longest distance from a vertex to a vertex. That is, for a pair of vertices $x, y \in V$, let $\mathrm{dist}(x, y)$ be the length of a shortest x-y (directed) path, and the *diameter* of T is:

$$\mathrm{diam}(T) := \max_{x,y \in V} \mathrm{dist}(x, y).$$

The following property is useful.

Lemma 1. *A tournament $T = (V, A)$ is efficient if and only if the diameter of T is less than or equal to 2.*

Proof. Let x and y be an arbitrary pair of vertices in V and we assume $(x, y) \in A$ w.l.o.g. Clearly $\mathrm{dist}(x, y) = 1$ and x is not defeated by y. If $\mathrm{dist}(y, x) \leq 2$, then there is a vertex $z \in V$ such that $(y, z), (z, x) \in A$. This means that y is not defeated by x. If $\mathrm{dist}(y, x) > 2$, then there is no vertex $z \in V$ such that $(y, z), (z, x) \in A$. This means that y is defeated by x and hence y is useless. From these discussions, the desired statement is directly obtained. □

By using this lemma, the case of $n = 4$ is easily checked as follows:

Assume that there is an efficient 4-tournament $T = (V, A)$. Let $(x, y) \in A$ be an arbitrary arc in A. From $\mathrm{diam}(T) \leq 2$, then $\mathrm{dist}(y, x) = 2$, i.e., there is a vertex $z \in V$ such that $(y, z), (z, x) \in A$. This means $\langle x, y, z, x \rangle$ is a 3-cycle. Let $w \in V$ be the remaining vertex. The out-degree and in-degree of w must be more than zero, since otherwise w is useless or superior than another vertex. Then there are a vertex that is defeated by w and a vetex that defeats w; w.l.o.g. we can let them x and y, respectively. The remaining freedom is only whether $(w, z) \in A$ or $(z, w) \in A$. In the former case, w is superior than z, and otherwise z is superior than w, i.e., T is not efficient, a contradiction. Therefore there is no efficient 4-tournament.

Now we consider the existence in some cases. For $n = 1$, the unique (trivial) 1-tournament ($V = \{x\}, A = \emptyset$) is efficient. (Note that this janken is in fact "useless" practically since in every time it ends in a tie, but it doesn't violate the condition of efficiency (Lemma 1), i.e., it includes no pair of signs in which one is superior than the other.) For $n = 3$, we know an efficient tournament as "rock-paper-scissors." For $n = 5$, the tournament of Fig. 2 (b) is efficient, you can check it doesn't have any useless sign.

We so far checked that for $n = 1, 3, 5$ there is an efficient tournament, and for $n = 2, 4$ there is not. But don't cursorily decide that if n is even number, we don't have efficient tournaments. For $n = 6$, we have three distinct efficient 6-tournaments as shown in Fig. 3. Except these three, we don't have any efficient 6-tournaments. It is not difficult to check it by using Lemma 1, and hence we omit to explain it.

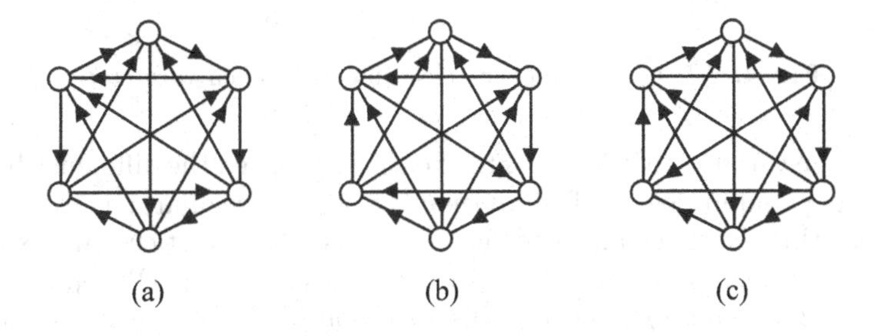

Fig. 3. Three efficient 6-jankens

For extending these results for general n, we use the following lemma.

Lemma 2. *Let $T_n = (V, A)$ be an efficient n-tournament, then the $(n+2)$-tournament $T_{n+2} = (V', A')$ defined below is also efficient.*

$$V' := V \cup \{x, y\}, \text{ where } x, y \notin V, \text{ and}$$
$$A' := A \cup \{(x, y)\} \cup \{(z, x), (y, z) \mid \forall z \in V\}.$$

Proof. For any $z \in V$, there is a 3-cycle $\langle z, x, y, z \rangle$. It follows from this that $\text{dist}(z, x) = \text{dist}(x, y) = \text{dist}(y, z) = 1$ and $\text{dist}(z, y) = \text{dist}(y, x) = \text{dist}(x, z) = 2$. By considering this and $\text{diam}(T_n) \leq 2$, we obtain that $\text{diam}(T_{n+2}) = 2$. □

Now we can prove Theorem 1:

Proof of Theorem 1: We observed that for $n = 1, 3, 5, 6$, there are efficient n-tournaments. By using Lemma 2, this theorem is obtained by induction. □

3 What Janken Is Amusing?

3.1 A Measure of Amusement

We proved that we can construct an efficient n-tournament for any positive integer n except 2 and 4. In fact for $n \geq 5$, there are more than one n-efficient tournaments. But now we have the next question: "Are they all the same for play?" For example, two 7-tournaments shown in Fig. 4 are both efficient. However they give a different impression if we play them.

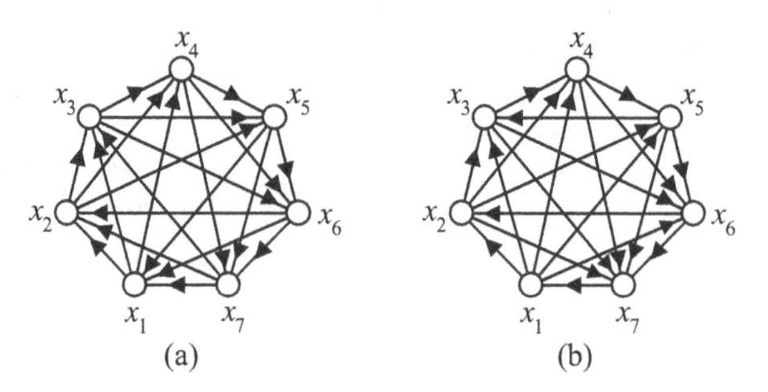

Fig. 4. What is different in these efficient janken variants?

Observe the variation of the strength of each vertex, i.e., the difference between the out-degree and in-degree. The tournament of (a) is regular, i.e., every vertex defeats just three vertices and is defeated by just three vertices (we express it as (3-3)). But in the tournament of (b), x_1 is (5-1), x_2 is (4-2), x_3, x_4, x_5 are (3-3), x_6 is (2-4), and x_7 is (1-5). This variation of the strength should effect the amusement of janken variants deeply. For example, in the janken variant of (a), every vertex is essentially the same and our strategy becomes very simple. However in the janken variant of (b), we consider that to use the strongest vertex x_1 have an advantage, but the opponent may gamble to use x_7, which is the weakest vertex but only one that defeats x_1. If you win by using x_7, you should feel very happy! Many games played widely employ such difference of strength, e.g., Napoleon, poker, contract bridge, war, gunjin-shogi (army chess), etc.

From this observation, we introduce a measure of amusement of tournaments (jankens) by using the variation of the difference of out-degree and in-degrees as follows.

$$\mathrm{irr}(T) := \sum_{x \in V} (\deg^+(x) - \deg^-(x))^2. \tag{1}$$

3.2 Proof of Theorem 2

Tournaments made by applying Lemma 2 one by one starting from the trivial 1-tournament look having the highest irregularity (amusement). Such tournaments have already defined and investigated by J.W. Moon [4]. Let we call them *Moon tournaments* and let M_n be the sets of Moon tournaments consists of n vertices, i.e., M_n is defined as follows:

Definition 2. *M_1 consists of the trivial 1-tournament. $M_2 = M_4 = \emptyset$. M_6 consists of the three efficient 6-tournaments shown in Fig. 3. For the other $n \in \{3, 5, 7, 8, 9, 10, \ldots\}$, M_n consists of T_n, which is obtained by applying the construction way of Lemma 2 to $T_{n-2} \in M_{n-2}$.*

From the definition, M_n consists of only one tournament if n is odd, and three tournaments if n is even except 2 and 4. Note that the popular 3-janken and the 7-janken of Fig. 4 (b) are the unique tournaments of M_3 and M_7, respectively.

For Moon tournaments the following result have been known.

Theorem 3 (Moon 93 [4]). *Let $c_3(T)$ be the number of 3-cycles in a tournament T. For an n-tournament T, if every arc is contained in a 3-cycle, then*

$$c_3(T) \geq \begin{cases} (n-1)^2/4, & \text{if } n \text{ is odd,} \\ (n^2 - 2n + 8)/4, & \text{if } n \text{ is even,} \end{cases}$$

with equality holding if $T \in M_n$.

To connect $c_3(T)$ and $\mathrm{irr}(T)$, we obtain the following property.

Lemma 3. *Let T be a tournament with $n \geq 1$ vertices. Then*

$$\mathrm{irr}(T) = \frac{n(n+1)(n-1)}{3} - 8c_3(T).$$

Proof. Let $S_n = (V_n, A_n)$ be the unique acyclic n-tournament, i.e., there is a linear order $\sigma : V_n \to \{1, 2, \ldots, n\}$ such that arc (x, y) exists if $\sigma(x) < \sigma(y)$. The following two equations clearly hold.

$$c_3(S_n) = 0, \tag{2}$$

$$\mathrm{irr}(S_n) = \sum_{i=1}^{n} (n + 1 - 2i)^2 = \frac{n(n+1)(n-1)}{3}. \tag{3}$$

Any n-tournament can be obtained by flipping arcs of S_n. Focus on an arbitrary arc (x, y) of an arbitrary n-tournament $T = (V, A)$. Let

$$\begin{aligned} k &:= |\{z \in V - \{x, y\} \mid (x, z), (z, y) \in A\}|, \\ h &:= |\{z \in V - \{x, y\} \mid (z, x), (y, z) \in A\}|, \\ p &:= |\{z \in V - \{x, y\} \mid (x, z), (y, z) \in A\}|, \text{ and} \\ q &:= |\{z \in V - \{x, y\} \mid (z, x), (z, y) \in A\}|. \end{aligned}$$

Let $T' = (V, A')$ be the tournament obtained by flipping arc (x, y) of T, i.e., $A' := A \cup \{(y, x)\} - \{(x, y)\}$.

$$\begin{aligned} \mathrm{irr}(T) - \mathrm{irr}(T') &= \left\{(k+p-h-q+1)^2 + (h+p-k-q-1)^2\right\} \\ &\quad - \left\{(k+p-h-q-1)^2 + (h+p-k-q+1)^2\right\} \\ &= 8(k-h) = -8(c_3(T) - c_3(T')). \end{aligned} \tag{4}$$

For any n-tournament T, there is a a sequence of n-tournaments $\langle T^0 = S_n, T^1, \ldots, T^k = T \rangle$, which transforms $T^0 = S_n$ into $T^k = T$, such that T^i is obtained from T^{i-1} by flipping one arc ($\forall i \in \{1, \ldots, k\}$). From equations (2), (3), and (4), we get:

$$\begin{aligned} \mathrm{irr}(T^k) - \mathrm{irr}(T^0) &= \sum_{i=1}^{k} \left(\mathrm{irr}(T^i) - \mathrm{irr}(T^{i-1})\right) = -8 \sum_{i=1}^{k} \left(c_3(T^i) - c_3(T^{i-1})\right) \\ &= -8 \left(c_3(T^k) - c_3(T^0)\right). \end{aligned}$$

Hence

$$\mathrm{irr}(T) = \mathrm{irr}(S_n) - 8c_3(T),$$

and the lemma is obtained. □

Now we can prove Theorem 2, which is presented in Section 1:

Proof of Theorem 2: Let T be an efficient n-tournament. From Lemma 1, every arc in T is contained in a 3-cycle. Hence T satisfies the inequality of Theorem 3. By considering Lemma 3, we get the following inequalities:

$$\mathrm{irr}(T) \leq \begin{cases} n(n+1)(n-1)/3 - 2(n-1)^2, & \text{if } n \text{ is odd}, \\ n(n+1)(n-1)/3 - 2(n^2 - 2n + 8), & \text{if } n \text{ is even}. \end{cases}$$

By calculating them, we obtain the inequalities in Theorem 2. □

3.3 Best Amusing Efficient 5-Janken

Based on these results, we introduce a new janken variant called "King-flea-janken" as follows. We use two extra signs "king" (only thumb extended) and "flea" (only little finger extended), whose roles are (see Fig. 5):

- *King* defeats all of rock, paper, and scissors, but is only defeated by flea.
- *Flea* is defeated by all of rock, paper, and scissors, but only defeats king.

From Theorem 2, it follows that king-flea-janken is the unique janken variant having the highest irregularity (i.e., amusement) among efficient 5-jankens. Although this janken is very simple, surprisingly we don't know that any 5-janken isomorphic to this have been presented.

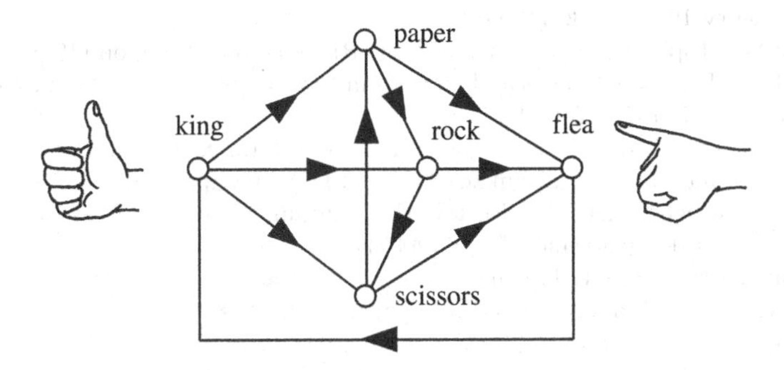

Fig. 5. King-flea-janken, which is the best amusing janken variant with five signs

We can easily calculate that the optimal mixed strategy (mixed Nash equilibrium) [2,6] of King-flea-janken is forming at random king, flea, rock, paper, and scissors with probability $1/3$, $1/3$, $1/9$, $1/9$, and $1/9$, respectively. It is interesting that the probability of the "weakest" sign flea and the "strongest" sign king are the same.

4 Conclusions

We considered generalizing janken in this paper. We defined "efficiency" of janken variants and introduced "a measure of amusement" of janken variants. Based on them, we first showed that we can construct an efficient janken variant with n signs for any positive integer n except 2 and 4. We next presented how to construct the best amusing janken variants with n signs, and gave "King-flea-janken."

We didn't treat any tie in janken variants. However there is some janken variants that allow ties between different signs (e.g., "god-rifle-fox-hen-termite" in Guangdong, China, some pairs, e.g., god and fox, end in ties. [5]). Extending our term "efficiency" and "the measure of amusement" to such janken variants with ties is remaining for future work.

Acknowledgements. We would like to express our gratitude to Prof. Hiroshi Maehara and Prof. Toshinori Sakai in Tokai University, Prof. Stefan Langerman in Université Libre de Bruxelles, and Prof. Erik Demaine in MIT for valuable information on janken variants. We would like to thank Prof. Yoshimi Egawa in Tokyo University of Science, Prof. Hikoe Enomoto in Keio University, and Prof. Hiroshi Nagamochi in Kyoto University for their precious discussions. We would also thank the anonymous referees for their careful reviews.

References

1. Chartrand, G., Lesniak, L., Zhang, P.: Graphs & Digraphs, 5th edn. CRC Press (2011)
2. Fisher, D.C., Ryan, J.: Tournament games and positive tournaments. Journal of Graph Theory 19, 217–236 (1995)
3. Moon, J.W.: Topics on Tournaments. Holt, Rinehart and Winston (1968)
4. Moon, J.W.: Uncovered nodes and 3-cycles in tournaments. Australasian Journal of Combinatorics 7, 157–173 (1993)
5. Ohbayashi, T., Kishino, U., Sougawa, T., Yamashita, S. (eds.): Encyclopedia of Ethnic Play and Games. Taishukan Shoten (1998) (in Japanese)
6. Nisan, N., Roughgarden, T., Tardos, E., Vazirani, V. (eds.): Algorithmic Game Theory. Cambridge University Press (2007)
7. Rock-paper-scissors, in Wikipedia, `http://en.wikipedia.org/wiki/Rock-paper-scissors`

Spanning Caterpillars Having at Most k Leaves

Mikio Kano[1,*], Tomoki Yamashita[2], and Zheng Yan[1,**]

[1] Department of Computer and Information Sciences
Ibaraki University, Hitachi, Ibaraki, Japan
kano@mx.ibaraki.ac.jp, yanzhenghubei@163.com
http://gorogoro.cis.ibaraki.ac.jp
[2] Department of Mathematics
Kinki University, Higashi-osaka, Osaka, Japan
yamashita@math.kindai.ac.jp

Abstract. A tree is called a caterpillar if all its leaves are adjacent to the same its path, and the path is called a spine of the caterpillar. Broersma and Tuinstra proved that if a connected graph G satisfies $\sigma_2(G) \geq |G| - k + 1$ for an integer $k \geq 2$, then G has a spanning tree having at most k leaves. In this paper we improve this result as follows. If a connected graph G satisfies $\sigma_2(G) \geq |G| - k + 1$ and $|G| \geq 3k - 10$ for an integer $k \geq 2$, then G has a spanning caterpillar having at most k leaves. Moreover, if $|G| \geq 3k - 7$, then for any longest path, G has a spanning caterpillar having at most k leaves such that its spine is the longest path. These three lower bounds on $\sigma_2(G)$ and $|G|$ are sharp.

1 Introduction

We consider simple graphs, which have neither loops nor multiple edges. For a graph G, let $V(G)$ and $E(G)$ denote the set of vertices and the set of edges of G, respectively. We write $|G|$ for the order of G (i.e., $|G| = |V(G)|$). For a vertex v of G, we denote by $\deg_G(v)$ the degree of v in G. We define $\sigma_2(G)$ to be the minimum degree sum of two nonadjacent vertices of G. An end-vertex of a tree is often called a *leaf*. A tree T is called a *caterpillar* if T contains a path such that all the vertices not contained in the path are adjacent to the path. In other words, a tree is a caterpillar if the removal of its leaves results in a path. Let T be a caterpillar. Then T has a path P connecting two leaves such that all the leaves of T not contained in P are adjacent to P. This path P is called a *spine* of T. Notice that the path Q obtained from T by removing all the leaves of T is often called the spine, however, for convenience, in this paper a spine of a caterpillar connects two leaves of a caterpillar and includes the path Q.

Recall the classic theorem of Ore [6] on a hamiltonian cycle.

Theorem 1 (Ore [6]). *Let G be a connected graph. If $\sigma_2(G) \geq |G|$, then G has a hamiltonian cycle.*

* Partially supported by Japan Society for the Promotion of Science, Grant-in-Aid for Scientific Research (C).

** Corresponding author.

J. Akiyama, M. Kano, and T. Sakai (Eds.): TJJCCGG 2012, LNCS 8296, pp. 95–100, 2013.

This theorem implies the following corollary on a hamiltonian path.

Corollary 1. *Let G be a connected graph. If $\sigma_2(G) \geq |G| - 1$, then G has a hamiltonian path.*

This corollary was generalized as follows by introducing a tree having at most k leaves. Notice that a hamiltonian path is a spanning tree with two leaves.

Theorem 2 (Broersma and Tuinstra [3]). *Let $k \geq 2$ be an integer and G be a connected graph. If $\sigma_2(G) \geq |G| - k + 1$, then G has a spanning tree with at most k leaves.*

Our main result is the following theorem, which says that under the same condition of Theorem 2, if the order of G is sufficiently large, then G has a spanning caterpillar having at most k leaves.

Theorem 3. *Let $k \geq 2$ be an integer and G be a connected graph. If $\sigma_2(G) \geq |G| - k + 1$ and $|G| \geq 3k - 10$, then G has a spanning caterpillar having at most k leaves.*

Furthermore, we obtain the following result, which requires the spine of a spanning caterpillar to be a given longest path.

Theorem 4. *Let $k \geq 2$ be an integer and let G be a connected graph. Let P be a longest path of G. If $\sigma_2(G) \geq |G| - k + 1$ and $|G| \geq 3k - 7$, G has a spanning caterpillar having at most k leaves such that its spine is P.*

We first show that the degree conditions of Theorems 3 and 4 are sharp. It is shown in [3] that the condition $\sigma_2(G) \geq |G| - k + 1$ is sharp for a graph to have a spanning tree with k leaves.

We now show that the order condition of Theorem 3 is sharp. Let K_m denote the complete graph of order m. Assume that $k \geq 6$. For each $1 \leq i \leq 3$, let H_i be a copy of K_{k-5}. We construct a graph G as follows: $V(G) = \{w, v_1, v_2, v_3\} \cup V(H_1) \cup V(H_2) \cup V(H_3)$ (disjoint union), w is adjacent all the vertices of $H_1 \cup H_2 \cup H_3$ and v_i is adjacent to all the vertices of H_i for each $1 \leq i \leq 3$. Then $|G| = 3(k-5) + 4 = 3k - 11$ and

$$\sigma_2(G) = \deg_G(v_1) + \deg_G(v_2) = 2(k-5) = |G| - k + 1.$$

However G has no spanning caterpillar. Thus the condition $|G| \geq 3k - 10$ is sharp.

We next show that the order condition of Theorem 4 is sharp. Assume that $k \geq 5$. Let H_i be a copy of K_{k-3} for each $i \in \{1, 2\}$ and let H_3 be a copy of K_{k-4}. We construct a graph G as follows: $V(G) = \{w, v_3\} \cup V(H_1) \cup V(H_2) \cup V(H_3)$ (disjoint union), w is adjacent all the vertices of $H_1 \cup H_2 \cup H_3$ and v_3 is adjacent to all the vertices of H_3. Then $|G| = 2(k-3) + (k-4) + 2 = 3k - 8$ and

$$\sigma_2(G) = \deg_G(v_3) + \deg_G(v) = k - 4 + k - 3 = |G| - k + 1,$$

where $v \in V(H_1) \cup V(H_2)$. However for a longest path P containing all vertices of $V(H_1) \cup V(H_2) \cup \{w\}$, G has no spanning caterpillar whose spine is P. Thus the condition $|G| \geq 3k - 7$ is sharp.

Czygrinow, Fan, Hurlbert, Kierstead and Trotter [4] investigated a spanning caterpillar with bounded degree in the same direction.

Another results on spanning trees with at most k leaves can be found in [5], [8] and others. The interested reader is referred to the survey paper [7] and the book [1] for more information on spanning trees.

2 Proof of Theorem 3

In this section, we give a proof of Theorem 3. Our proof uses the following result on dominating paths of graphs. For a graph G, let $\sigma_3(G)$ is defined to be the minimum degree sum of three independent vertices of G, where a vertex set X is called independent if no two vertices of X are adjacent in G.

Lemma 1 (Broersma [2], Corollary 14 ($k = 1$ and $\lambda = 2$)). *Let G be a connected graph. If $\sigma_3(G) \geq |G| - 3$, then G has a spanning caterpillar.*

Proof of Theorem 3. Let $\{x, y, z\}$ be any set of three independent vertices of G. Then

$$\begin{aligned}
\sigma_3(G) &\geq \deg_G(x) + \deg_G(y) + \deg_G(z) \\
&= \frac{\deg_G(x) + \deg_G(y)}{2} + \frac{\deg_G(y) + \deg_G(z)}{2} + \frac{\deg_G(z) + \deg_G(x)}{2} \\
&\geq \frac{3\sigma_2(G)}{2} \geq \frac{3(|G| - k + 1)}{2} \\
&\geq |G| - \frac{7}{2}. \qquad \text{(by } |G| \geq 3k - 10\text{)}
\end{aligned}$$

Hence by Lemma 1, G has a spanning caterpillar.

Choose a spanning caterpillar T of G so that its spine is as long as possible. Let P be a spine of T, and let u and v be the two end-vertices of P, which are leaves of T. We assign an orientation in P from u to v, and for a vertex x of P, its *successor* x^+ and the *predecessor* x^- are defined, if they exist. By the choice of the spanning caterpillar, G has no cycle C with $V(C) = V(P)$, and it follows that $N_G(u) \cap (V(G) - V(P)) = \emptyset$, $N_G(v) \cap (V(G) - V(P)) = \emptyset$ and $N_G(u)^- \cap N_G(v) = \emptyset$. Since $N_G(u)^- \cup N_G(v) \subseteq V(P) - \{v\}$, we obtain

$$\deg_G(u) + \deg_G(v) \leq |P| - 1.$$

Since $\sigma_2(G) \geq |G| - k + 1$, we have $|G| - k + 1 \leq |P| - 1$, which implies $|G| - |P| \leq k - 2$. Therefore the spanning caterpillar T has at most k leaves. □

3 Proof of Theorem 4

In this section, we give a proof of Theorem 4. We denote by $P[u, v]$ a path connecting two vertices u and v, which are the end-vertices of P. For a vertex set X of a graph G, let $\langle X \rangle_G$ denote the subgraph of G induced by X.

Proof of Theorem 4. If G has a hamiltonian path, we are done, and so we may assume that G does not have a hamiltonian path. Let P be a longest path in G, and let u and v be the end-vertices of P. We assign an orientation in P from u to v, and for a vertex x of P, we denote its successor and predecessor, if any, by x^+ and x^-, respectively. The following claim holds immediately by the fact that P is a longest path of G.

Claim 1. (i) $N_G(u) \cup N_G(v) \subseteq V(P)$.
(ii) G has no cycle C with $V(C) = V(P)$.
(iii) $N_G(u)^- \cap N_G(v) = \emptyset$ and $\{v\} \cup N_G(u)^- \cup N_G(v) \subseteq V(P)$.

By Claim 1, we have

$$\begin{aligned} |V(P)| &\geq |N_G(u)^-| + |N_G(v)| + |\{v\}| = \deg_G(u) + \deg_G(v) + 1 \\ &\geq \sigma_2(G) + 1 \geq |G| - k + 2. \end{aligned} \tag{1}$$

Hence $|G| - |V(P)| \leq k - 2$. Since G is a connected graph, by connecting all the vertices in $V(G) - V(P)$ to P by edges or paths, we can obtain a spanning tree T of G with at most k leaves.

Next, we prove that T is a caterpillar. Otherwise, there exists a vertex $w \in V(G) - V(P)$ such that $N_G(w) \cap V(P) = \emptyset$. By the choice of w and by Claim 1, the following claim easily holds.

Claim 2. (i) $\{w, u, v\}$ is an independent set of G.
(ii) $N_G(w) \subseteq V(G) - V(P) - \{w\}$.

By Claim 2 (i), we have

$$\deg_G(w) + \deg_G(u) + \deg_G(v) \geq \frac{3\sigma_2(G)}{2} \geq \frac{3}{2}(|G| - k + 1). \tag{2}$$

On the other hand, it follows from Claim 2 (ii) and Claim 1 that

$$\begin{aligned} &\deg_G(w) + \deg_G(u) + \deg_G(v) \\ &= |N_G(w)| + |N_G(u)^-| + |N_G(v)| \\ &\leq |G| - |P| - 1 + |P| - 1 = |G| - 2. \end{aligned} \tag{3}$$

By (2) and (3), we have $|G| \leq 3k-7$. Hence, the theorem holds when $|G| \geq 3k-6$.

Next we consider the case where $|G| = 3k-7$. In this case, $\sigma_2(G) \geq |G|-k+1 = 2k-6$. Furthermore, if $k \leq 4$, then $|G| \leq 5$ and so the theorem holds. Hence we may assume that $k \geq 5$.

Assume that $|P| = 2k - 5 + t$, where $t \geq 0$ by (1). Then $\deg_G(w) \leq |G| - |P| - 1 = k - 3 - t$. Hence,

$$\deg_G(w) + \min\{\deg_G(u), \deg_G(v)\} \leq k - 3 - t + \frac{|P| - 1}{2} = 2k - 6 - \frac{t}{2}.$$

Since $\sigma_2(G) \geq 2k - 6$, we obtain $t = 0$, that is, $|P| = 2k - 5$. Since the above inequality holds with equality, it follows from (1) that $\deg_G(w) = \deg_G(u) = \deg_G(v) = k - 3$. Since $\sigma_2(G) \geq 2k - 6$, we have

$$\deg_G(x) \geq k - 3 \qquad \text{for every vertex} \quad x \in V(G) - V(P). \tag{4}$$

Since the inequality (1) holds with equality,

$$V(P) - \{v\} = N_G(u)^- \cup N_G(v) \quad \text{(disjoint union).} \tag{5}$$

Since G is connected, G has a path Q connecting w and a vertex of $V(P)$. Note that Q has at least two vertices. Let $\{z\} = V(P) \cap V(Q)$. Since P is a longest path, $z \notin N_G(u)^-$. By (5), we obtain $z \in N_G(v)$. Since P is longest, we have $z^+ \notin N_G(u)^-$; otherwise $Q[w, z] + P[z, u] + uz^{++} + P[z^{++}, v]$ is a longer path than P. Hence $z^+ \in N_G(v)$. Inductively by using (5) and Calim 1, we obtain that $s \in N_G(v)$ for every vertex $s \in V(P[z, v^-])$. It is immediate that $z^- \notin N_G(v)$, which implies $z^- \in N_G(u)^-$ by (5), and thus $z \in N_G(u)$. Inductively, we can show that $t \in N_G(u)$ for every vertex $t \in V(P[u^+, z])$. By the fact that P is a longest path of G, for every vertex $x \in V(G) - V(P)$, it follows that $N_G(u)^- \cap N_G(x) = \emptyset$ and $N_G(v)^+ \cap N_G(x) = \emptyset$. Therefore we obtain

$$N_G(x) \cap V(P) \subseteq \{z\} \qquad \text{for every vertex} \quad x \in V(G) - V(P). \tag{6}$$

Claim 3. $H = \langle (V(G) - V(P)) \cup \{z\} \rangle_G$ has a hamiltonian path with an end-vertex z.

We prove Claim 3. If $\langle V(G) - V(P) \rangle_G$ is a complete graph, then we are done. We may assume that $\langle V(G) - V(P) \rangle_G$ is not complete. By (6), (4) and $k \geq 5$, we have $\deg_H(x) = \deg_G(x) \geq k - 3 \geq 2$ for each vertex $x \in V(G) - V(P)$.

Since $\langle V(G) - V(P) \rangle_G$ is not complete, there exists two non-adjacent vertices s and t in it, which are adjacent to z since $\deg_H(s) + \deg_H(t) \geq \sigma_2(G) \geq 2(|H| - 2)$, and hence $\deg_H(z) \geq 2$. Therefore $\deg_H(x) + \deg_H(y) \geq k - 3 + 2 = |H|$ for all non-adjacent two vertices $x, y \in V(H)$. By Theorem 1, H has a hamiltonian cycle, and so H has a hamiltonian path with end-vertex z. Therefore Claim 3 is proved.

Let R be a hamiltonian path with end-vertex z in H, and x be another end-vertex of R. Then $R[x, z] + P[z, u]$ or $R[x, z] + P[z, v]$ is a path of order at least $k - 2 + (|V(P)| + 1)/2 > |V(P)|$, a contradiction.

Consequently, Theorem 4 is proved. □

References

1. Akiyama, J., Kano, M.: Factors and Factorizations of Graphs. LNM, vol. 2031. Springer (2011)
2. Broersma, H.J.: Existence of Δ_λ-cycles and Δ_λ-paths. J. Graph Theory 12, 499–507 (1988)
3. Broersma, H.J., Tuinstra, H.: Independence trees and hamilton cycles. J. Graph Theory 29, 227–237 (1998)
4. Czygrinow, A., Fan, G., Hurlbert, G., Kierstead, H.A., Trotter, W.T.: Spanning trees of bounded degree. Electronic Journal of Combinatorics 8, R33 (2001)
5. Kano, M., Kyaw, A.: A note on leaf-constrained spanning trees in a graph. Ars Combinatoria 108, 321–326 (2013)
6. Ore, O.: Note on hamilton citcuits. Amer. Math. Monthly 67, 66 (1960)
7. Ozeki, K., Yamashita, T.: Spanning trees: A survey. Graphs Combinatorics 22, 1–26 (2011)
8. Win, S.: On a conjecture of Las Vergnas concerning certain spanning trees in graphs. Resultate Math. 2, 215–224 (1979)

GDDs with Two Associate Classes and with Three Groups of Sizes $1, n, n$ and $\lambda_1 < \lambda_2$

Wannee Lapchinda[1,*] Narong Punnim[1,**] and Nittiya Pabhapote[2,*,***]

[1] Srinakharinwirot University, Sukhumvit 23, Bangkok 10110, Thailand
gs522120002@swu.ac.th, narongp@swu.ac.th

[2] University of the Thai Chamber of Commerce, Bangkok 10400, Thailand
nittiya_pab@utcc.ac.th

Abstract. A group divisible design $\text{GDD}(v = 1 + n + n, 3, 3, \lambda_1, \lambda_2)$ is an ordered pair $(V, \mathcal{B})$ where V is an $(1 + n + n)$-set of symbols and $\mathcal{B}$ is a collection of 3-subsets (called blocks) of V satisfying the following properties: the $(1 + n + n)$-set is divided into 3 groups of sizes 1, n and n; each pair of symbols from the same group occurs in exactly λ_1 blocks in $\mathcal{B}$; and each pair of symbols from different groups occurs in exactly λ_2 blocks in $\mathcal{B}$. Let λ_1, λ_2 be positive integers. Then the spectrum of λ_1, λ_2, denoted by $\mathsf{Spec}(\lambda_1, \lambda_2)$, is defined by

$$\mathsf{Spec}(\lambda_1, \lambda_2) = \{n \in \mathbb{N} : a \ \text{GDD}(v = 1 + n + n, 3, 3, \lambda_1, \lambda_2) \text{ exists}\}.$$

We found in [10] the spectrum $\mathsf{Spec}(\lambda_1, \lambda_2)$ provided that $\lambda_1 \geq \lambda_2$ in all situations. We find in this paper $\mathsf{Spec}(\lambda_1, \lambda_2)$ when $\lambda_1 < \lambda_2$ in all situations.

1 Introduction

A *group divisible design* $\text{GDD}(v = v_1 + v_2 + \ldots + v_g, g, k, \lambda_1, \lambda_2)$ is a collection of k-subsets (called blocks) of a v-set of symbols, where the v-set is partitioned into g groups of sizes $v_1, v_2, \ldots, v_g$; each pair of symbols from the same group occurs in exactly λ_1 blocks; and each pair of symbols from different groups occurs in exactly λ_2 blocks. Elements occurring together in the same group are called *first associates*, and elements occurring in different groups are called *second associates*. The existence problem of such GDDs has been of interest over the years, going back to at least the work of Bose and Shimamoto in 1952 who began classifying such designs [1]. More recently, much work has been done on the existence of such designs when $\lambda_1 = 0$ (see [3] for a summary), and the designs here are called partially balanced incomplete block designs (PBIBDs) of group divisible type in [3]. The existence question for $k = 3$ has been solved by Sarvate, Fu and Rodger (see [4], [5]) when all groups are the same size.

* Supported by University of the Thai Chamber of Commerce.
** Supported by The Thailand Research Fund.
*** Corresponding author.

J. Akiyama, M. Kano, and T. Sakai (Eds.): TJJCCGG 2012, LNCS 8296, pp. 101–109, 2013.

The existence problem of $\mathrm{GDD}(v = v_1 + v_2 + \ldots + v_g, g, k, \lambda_1, \lambda_2)$, when the groups may have different size, is considered recently. Chaiyasena, et al. [2] have published a paper in this direction. In particular, they found all ordered pairs (n, λ) of positive integers such that a $\mathrm{GDD}(v = 1 + n, 2, 3, 1, \lambda)$ exists. Details can be found in [2]. Pabhapote and Punnim found in [12] all ordered triples (m, n, λ) of positive integers with such that a $\mathrm{GDD}(v = m + n, 2, 3, \lambda, 1)$ exists. The existence problem of a $\mathrm{GDD}(v = m + n, 2, 3, \lambda_1, \lambda_2)$ is more difficult if $\lambda_1 < \lambda_2$. Punnim and Uiyyasathian found in [11] infinitely many ordered pairs (m, n) of positive integers such that a $\mathrm{GDD}(v = m + n, 2, 3, 1, 2)$ exists.

We now consider the problem of determining the existence of a $\mathrm{GDD}(v = n_1 + n_2 + n_3, 3, 3, \lambda_1, \lambda_2)$. Chaiyasena, et al. [2] published a paper in this direction for small values of n_1, n_2, n_3. In particular, for each $n \in \{2, 3, 4, 5, 6\}$ they found all ordered pairs (λ_1, λ_2) of positive integers such that a $\mathrm{GDD}(v = 1 + 2 + n, 3, 3, \lambda_1, \lambda_2)$ exists. Hurd and Sarvate found in [6] all ordered pairs (n, λ) of positive integers such that a $\mathrm{GDD}(v = 1 + 1 + n, 3, 3, 1, \lambda)$ exists. Later, Hurd and Sarvate found in [7] all ordered pairs (n, λ) of positive integers such that a $\mathrm{GDD}(v = 1 + 1 + n, 3, 3, \lambda, 1)$ exists. Recently, Hurd and Sarvate found in [8] all ordered triples $(n, \lambda_1, \lambda_2)$ of positive integers, with $\lambda_1 > \lambda_2$, such that a $\mathrm{GDD}(v = 1 + 2 + n, 3, 3, \lambda_1, \lambda_2)$ exists.

We will concentrate in this paper on GDDs with three groups of sizes $1, n$ and n and consider the following problem:

Problem: Find all triples $(n, \lambda_1, \lambda_2)$ of positive integers such that there exists a $\mathrm{GDD}(v = 1 + n + n, 3, 3, \lambda_1, \lambda_2)$.

This problem was recently solved in [10] when $\lambda_1 \geq \lambda_2$. Thus, this paper is a continuation of what we have done in [10]. Let λ_1, λ_2 be positive integers with $\lambda_1 < \lambda_2$. The spectrum of λ_1, λ_2, denoted by $\mathsf{Spec}(\lambda_1, \lambda_2)$, is defined by

$$\mathsf{Spec}(\lambda_1, \lambda_2) = \{n \in \mathbb{N} : a\ \mathrm{GDD}(v = 1 + n + n, 3, 3, \lambda_1, \lambda_2) \text{ exists}\}.$$

We proved in [10] the following theorem.

Theorem 1. *Let n be a positive integer. If $n \in \mathsf{Spec}(\lambda_1, \lambda_2)$, then n is a solution of the system of linear congruences:*

$$F(\lambda_1, \lambda_2) = \lambda_1 n(n-1) + \lambda_2 n(n+2) \equiv 0 (\mathrm{mod}\ 3) \quad \cdots \quad (1)$$
$$G(\lambda_1, \lambda_2) = \lambda_1 (n-1) + \lambda_2 (n+1) \equiv 0\ (\mathrm{mod}\ 2) \cdots \quad (2).$$

The following table shows the relationship between n (mod 6) and (λ_1, λ_2) as given in Theorem 1.

In the case where $\lambda_1 < \lambda_2$ we have an additional necessary condition.

Lemma 1. *If* $\mathrm{GDD}(v = 1 + n + n, 3, 3, \lambda_1, \lambda_2)$ *exists, then it is necessary that* $\lambda_2 \leq 2\lambda_1$.

Proof. Let $X = \{x\}$, $Y = \{y_1, y_2, \ldots, y_n\}$, and $Z = \{z_1, z_2, \ldots, z_n\}$. Let $P = X \cup Y$. Thus there are $\lambda_2 n(n+1)$ edges between P and Z, there are $\lambda_1 \binom{n}{2}$ edges inside Z, and there are $\lambda_1 \binom{n}{2} + \lambda_2 n$ edges inside P. Since any three edges

Table 1. Necessity

λ_1 \ λ_2	0	1	2	3	4	5
0	all n	1, 3	0, 1, 3, 4	1, 3, 5	0, 1, 3, 4	1, 3
1	1,3	0, 1, 3, 4	1, 3, 5	0, 1, 3, 4	1, 3	all n
2	0, 1, 3, 4	1, 3, 5	0, 1, 3, 4	1, 3	all n	1, 3
3	1, 3, 5	0, 1, 3, 4	1, 3	all n	1, 3	0, 1, 3, 4
4	0, 1, 3, 4	1, 3	all n	1, 3	0, 1, 3, 4	1, 3, 5
5	1, 3	all n	1, 3	0, 1, 3, 4	1, 3, 5	0, 1, 3, 4

between P and Z can not form a triangle, and any one edge in P and any one edge in Z can also not form a triangle, it follows that triangles can be formed by three edged in P, three edges in Z, or two edges between P and Z and one edge in $P \cup Z$. Thus, the number of edges between P and Z can not exceed twice the number of edges in $P \cup Z$. Therefore,

$$\lambda_2 n(n+1) \leq 2[\lambda_1 \binom{n}{2} + \lambda_1 \binom{n}{2} + \lambda_2 n].$$

That is $\lambda_2 \leq 2\lambda_1$.

2 Preliminary Results

We review some known results concerning triple designs that will be used in what follows. Most of these results are taken from [9]. We begin with the classic well-known result on the existence of balanced incomplete block designs (BIBDs).

Theorem 2. *Let v be a positive integer. Then there exists a* BIBD$(v, 3, 1)$ *if and only if* $v \equiv 1$ *or* $3 \pmod 6$.

A BIBD$(v, 3, 1)$ is usually called *Steiner triple system* and is denoted by STS(v). Let $(V, \mathcal{B})$ be an STS(v). Then the number of triples $b = |\mathcal{B}| = v(v-1)/6$. A *parallel class* in an STS(v) is a set of disjoint triples whose union is V. A parallel class contains $v/3$ triples, and hence an STS(v) having a parallel class can exist only when $v \equiv 3 \pmod 6$. When the set $\mathcal{B}$ can be partitioned into parallel classes, such a partition $\mathcal{R}$ is called a *resolution* of the STS(v), and the STS(v) is called *resolvable.* If $(V, \mathcal{B})$ is an STS(v) and $\mathcal{R}$ is a resolution of it, then $(V, \mathcal{B}, \mathcal{R})$ is called a *Kirkman triple system*, denoted by KTS(v), with $(V, \mathcal{B})$ as its *underlying* STS. It is well known that a KTS(v) exists if and only if $v \equiv 3 \pmod 6$. Thus if $(V, \mathcal{B}, \mathcal{R})$ is a KTS(v), then $\mathcal{R}$ contains $(v-1)/2$ parallel classes. The following results on the existence of λ-fold triple systems are well known (see, e.g., [9]).

Theorem 3. *Let v be a positive integer. Then a* BIBD$(v, 3, \lambda)$ *exists if and only if λ and v are in one of the following cases:*

(a) $\lambda \equiv 0 \pmod 6$ *and* $v \neq 2$,
(b) $\lambda \equiv 1$ *or* $5 \pmod 6$ *and* $v \equiv 1$ *or* $3 \pmod 6$,
(c) $\lambda \equiv 2$ *or* $4 \pmod 6$ *and* $v \equiv 0$ *or* $1 \pmod 3$, *and*
(d) $\lambda \equiv 3 \pmod 6$ *and* v *is odd.*

A *factor* of a graph G is a spanning subgraph. An r*-factor* of a graph is a spanning r-regular subgraph, and an r*-factorization* is a partition of the edges of the graph into disjoint r-factors. A graph G is said to be r*-factorable* if it admits an r-factorization. In particular, a 1-factor is a *perfect matching,* and a 1-factorization of an r-regular graph G is a set of 1-factors which partition the edge set of G.

A *near* 1*-factorization* of a regular graph on $2x+1$ vertices is a partition of its edges into independent sets of size x, each of which is called a *near* 1*-factor*. Each near 1-factor saturates all vertices except one; the exceptional vertex is known as its *deficiency.* A near 1-factorization of K_{2x+1} can be constructed from a 1-factorization of K_{2x+2} by deleting a single vertex.

The following notations will be used throughout the paper for our constructions.

1. Let $T = \{x, y, z\}$ be a triple and $a \notin T$. We use $a * T$ for three triples of the form $\{a, x, y\}, \{a, x, z\}, \{a, y, z\}$. If $\mathcal{T}$ is a set of triples, then $a * \mathcal{T}$ is defined as $\{a * T : T \in \mathcal{T}\}$.
2. Let $G = \langle V(G), E(G)\rangle$ and $H = \langle V(H), E(H)\rangle$ be two vertex disjoint simple graphs. If $e = uv \in E(G)$ and $a \in V(H)$, then we use $a + e$ for the triple $\{a, u, v\}$. If $\emptyset \neq X \subseteq E(G)$, then we use $a + X$ for the collection of triples $a + e$ for all $e \in X$.
3. Let V be a v-set. We use $K(V)$ for the complete graph K_v on the vertex set V.
4. Let V be a v-set. BIBD$(V, 3, \lambda)$ can be defined as BIBD$(V, 3, \lambda) = \{\mathcal{B} : (V, \mathcal{B})$ is a BIBD$(v, 3, \lambda)\}$.
5. Let X and Y be disjoint sets of cardinality m and n, respectively. We define GDD$(X, Y; \lambda_1, \lambda_2)$ as GDD$(X, Y; \lambda_1, \lambda_2) = \{\mathcal{B} : (X, Y; \mathcal{B})$ is a GDD$(v = m + n, 2, 3, \lambda_1, \lambda_2)\}$.
6. Let X, Y and Z be three pairwise disjoint sets of cardinality n_1, n_2 and n_3, respectively. We define GDD$(X, Y, Z; \lambda_1, \lambda_2)$ as GDD$(X, Y, Z; \lambda_1, \lambda_2) = \{\mathcal{B} : (X, Y, Z; \mathcal{B})$ is a GDD$(v = n_1 + n_2 + n_3, 3, 3, \lambda_1, \lambda_2)\}$.
7. When we say that $\mathcal{B}$ is a *collection* of subsets (blocks) of a v-set V, $\mathcal{B}$ may contain repeated blocks. Thus "$\cup$" in our context will be used for the union of multisets.
8. Finally, if we have a set X, the cardinality of X is denoted by $|X|$.

3 Sufficiency

We will assume throughout this subsection that $\lambda_1 < \lambda_2$. It is well known that K_n has a 1-factorization if and only if n is even. If $n = 2m + 1 \geq 3$ is odd, then K_n can be decomposed as a union of n copies of $mK_2 \cup \{v\}$. In this case

$F_v = mK_2$ is a near 1-factor and v is the deficiency of F_v. Let $X = \{x\}$, $Y = \{y_1, y_2, \ldots, y_n\}$ and $Z = \{z_1, z_2, \ldots, z_n\}$. Let $\mathcal{F} = \{F_{y_i} : i = 1, 2, \ldots, n\}$ and $\mathcal{F}' = \{F_{z_i} : i = 1, 2, \ldots, n\}$ be the sets of near 1-factorizations of $K(Y)$ and $K(Z)$, respectively. Let $\mathcal{B}_1 = \bigcup_{i=1}^{n} y_i + F_{z_i}$, $\mathcal{B}_2 = \bigcup_{i=1}^{n} z_i + F_{y_i}$, and $\mathcal{B}_3 = \{\{x, y_i, z_i\} : i = 1, 2, \ldots, n\}$. It can be easily checked that $(X, Y, Z; \mathcal{B}_1 \cup \mathcal{B}_2 \cup 2\mathcal{B}_3)$ forms a GDD$(v = 1+n+n, 3, 3, 1, 2)$ for odd integers n. Consequently, for every positive integer λ, GDD$(v = 1+n+n, 3, 3, \lambda, 2\lambda)$ exists. Therefore, we obtain the following lemma.

Lemma 2. *Let n be an odd positive integer. Then for every positive integer λ we have $n \in \mathsf{Spec}(\lambda, 2\lambda)$.*

Let $n \equiv 2 \pmod 6$ and let $X = \{x\}, Y = \{y_1, y_2, \ldots, y_n\}$ and $Z = \{z_1, z_2, \ldots, z_n\}$. Thus KTS$(X \cup Y)$ and KTS$(X \cup Z)$ are not empty. We now choose $\mathcal{B}_1 \in$ KTS$(X \cup Y)$ with $\mathcal{P}_1, \mathcal{P}_2, \ldots, \mathcal{P}_{\frac{n-1}{2}}$ as its parallel classes, and $\mathcal{B}_2 \in$ KTS$(X \cup Z)$ with $\mathcal{Q}_1, \mathcal{Q}_2, \ldots, \mathcal{Q}_{\frac{n-1}{2}}$ as its parallel classes. Let

$$\mathcal{B} = \left(\bigcup_{i=1}^{\frac{n-1}{2}} z_i * \mathcal{P}_i\right) \cup \left(\bigcup_{i=\frac{n+1}{2}}^{n} z_i * \mathcal{P}_{i-\frac{n-1}{2}}\right) \cup \left(\bigcup_{i=1}^{\frac{n-1}{2}} y_i * \mathcal{Q}_i\right) \cup \left(\bigcup_{i=\frac{n+1}{2}}^{n} y_i * \mathcal{Q}_{i-\frac{n-1}{2}}\right).$$

It can be easily checked that $(X, Y, Z; \mathcal{B})$ forms a GDD$(v = 1+n+n, 3, 3, 2, 4)$. Consequently, for every positive integer λ, GDD$(v = 1+n+n, 3, 3, 2\lambda, 4\lambda)$ exists. Therefore, we obtain the following lemma.

Lemma 3. *Let $n \equiv 2 \pmod 6$. Then for every positive integer λ we have $n \in \mathsf{Spec}(2\lambda, 4\lambda)$.*

Let $n \equiv 0$ or $4 \pmod 6$ and let $X = \{x\}$, $Y = \{y_1, y_2, \ldots, y_n\}$ and $Z = \{z_1, z_2, \ldots, z_n\}$. It follows, by Theorem 3, that BIBD$(Y, 3, 2)$ and BIBD$(Z, 3, 2)$ are not empty. Let $\mathcal{B}_1 \in$ BIBD$(Y, 3, 2)$. Then we choose the corresponding $\mathcal{B}_2 \in$ BIBD$(Z, 3, 2)$ in a natural way. Namely, $\{z_i, z_j, z_k\}$ is a block in $\mathcal{B}_2$ if and only if $\{y_i, y_j, y_k\}$ is a block in $\mathcal{B}_1$. Let $\mathcal{B}_3 = \{\{y_i, y_j, z_k\} : \{y_i, y_j, y_k\} \in \mathcal{B}_1\}$, $\mathcal{B}_4 = \{\{z_i, z_j, y_k\} : \{z_i, z_j, z_k\} \in \mathcal{B}_2\}$, and $\mathcal{B}_5 = \{\{x, y_i, z_i\} : i = 1, 2, \ldots, n\}$. Now let $\mathcal{B} = \mathcal{B}_3 \cup \mathcal{B}_4 \cup 4\mathcal{B}_5$. Then it can be checked that $(X, Y, Z; \mathcal{B})$ forms a GDD$(v = 1+n+n, 3, 3, 2, 4)$. Consequently, for every positive integer i, GDD$(v = 1+n+n, 3, 3, 2i, 4i)$ exists. Therefore, we obtain the following lemma.

Lemma 4. *Let $n \equiv 0$ or $4 \pmod 6$. Then for every positive integer λ we have $n \in \mathsf{Spec}(2\lambda, 4\lambda)$.*

Results in Lemmas 2-4 are our basic tools for constructing the GDDs in general and we can see that if n is odd, then GDD$(v = 1+n+n, 3, 3, \lambda, 2\lambda)$ exists for every positive integer λ and if n is even, then GDD$(v = 1+n+n, 3, 3, \lambda, 2\lambda)$ exists for every positive even integer λ. We can summarize as a basic theorem.

Theorem 4. *Let n be a positive integer. If n is odd, then* GDD$(v = 1+n+n, 3, 3, \lambda, 2\lambda)$ *exists for every positive integer λ and if n is even, then* GDD$(v = 1+n+n, 3, 3, \lambda, 2\lambda)$ *exists for every positive even integer λ.*

We will use in the following results X, Y, Z for sets of sizes $1, n, n$, respectively. We first observe that $2n+1 \equiv 1$ or $3 \pmod 6$ if and only if $n \equiv 0$, 1, 3 or $4 \pmod 6$. In these particular values of n, BIBD$(2n+1, 3, 1)$ exists, and hence, BIBD$(2n+1, 3, \lambda)$ exists for every positive integer λ. For $n \equiv 2$ or $5 \pmod 6$, we can ensure that BIBD$(2n+1, 3, 3\lambda)$ exists for every positive integer λ. Let $n \equiv 0$, 1, 3 or $4 \pmod 6$. Let $\mathcal{B}_1 \in$ BIBD$(X \cup Y \cup Z, 3, 1)$. Let s and t be non-negative integers and $i \in \{1,2,3,4,5,6\}$ such that $6t+i < 6s+i \leq 2(6t+i)$. Thus, $6(s-2t) \leq i$. Then, by Theorem 4, GDD$(v = 1+n+n, 3, 3, 6(s-t), 12(s-t))$ exists. Let $\mathcal{B}_2 \in$ GDD$(X, Y, Z; 6(s-t), 12(s-t))$. Thus, $(X, Y, Z; \mathcal{B})$ forms a GDD$(v = 1+n+n, 3, 3, 6t+i, 6s+i)$, where $\mathcal{B} = (6(2t-s)+i)\mathcal{B}_1 \cup \mathcal{B}_2$. Thus, we have the following result.

Lemma 5. *Let $n \equiv 0$, 1, 3 or $4 \pmod 6$ and $i \in \{1,2,3,4,5,6\}$. If s and t are non-negative integers with $6t+i < 6s+i \leq 2(6t+i)$, then $n \in \mathsf{Spec}(6t+i, 6s+i)$.*

Suppose that $n \equiv 2$ or $5 \pmod 6$ and $i \in \{3, 6\}$. If s and t are non-negative integers with $6t+i < 6s+i \leq 2(6t+i)$, then, by Theorem 3, BIBD$(2n+1, 3, 3)$ exists. Let $\mathcal{B}_1 \in$ BIBD$(X \cup Y \cup Z, 3, 3)$. By Theorem 4, GDD$(v = 1+n+n, 3, 3, 6(s-t), 12(s-t))$ exists. Let $\mathcal{B}_2 \in$ GDD$(X, Y, Z; 6(s-t), 12(s-t))$. It can be checked that $(X, Y, Z; \mathcal{B})$ forms a GDD$(v = 1+n+n, 3, 3, 6t+i, 6s+i)$, where $\mathcal{B} = (2(2t-s)+\frac{i}{3})\mathcal{B}_1 \cup \mathcal{B}_2$. Thus, we have the following results.

Lemma 6. *Let $n \equiv 2$ or $5 \pmod 6$ and $i \in \{3, 6\}$. If s and t are non-negative integers with $6t+i < 6s+i \leq 2(6t+i)$, then $n \in \mathsf{Spec}(6t+i, 6s+i)$.*

Note that the results of Lemma 5 and Lemma 6 take care of the sufficiency for the values of the entries on the main diagonal of Table 1.

Let $n \equiv 1$ or $3 \pmod 6$, s and t be non-negative integers, and $i \in \{1,2,3,4,5,6\}$ with $6t+i < 6s+i+1 \leq 2(6t+i)$. Then, by Theorem 4, let $\mathcal{B}_1 \in$ GDD$(X, Y, Z; 6(s-t)+1, 12(s-t)+2)$ and, by Theorem 3, let $\mathcal{B}_2 \in$ BIBD$(X \cup Y \cup Z, 3, 1)$. Thus, $(X, Y, Z; \mathcal{B})$ forms a GDD$(v = 1+n+n, 3, 3, 6t+i, 6s+i+1)$, where $\mathcal{B} = \mathcal{B}_1 \cup (6(2t-s)+i-1)\mathcal{B}_2$.

Let $n \equiv 5 \pmod 6$, s and t be non-negative integers, and $i \in \{1, 4\}$ with $6t+i < 6s+i+1 \leq 2(6t+i)$. Then, by Theorem 4, let $\mathcal{B}_1 \in$ GDD$(X, Y, Z; 6(s-t)+1, 12(s-t)+2)$ and, by Theorem 3, let $\mathcal{B}_2 \in$ BIBD$(X \cup Y \cup Z, 3, 3)$. Thus, $(X, Y, Z; \mathcal{B})$ forms a GDD$(v = 1+n+n, 3, 3, 6t+i, 6s+i+1)$, where $\mathcal{B} = \mathcal{B}_1 \cup (\frac{1}{3}(6(2t-s)+i-1))\mathcal{B}_2$. Thus, we have the following results.

Lemma 7. *Let $n \equiv 1$ or $3 \pmod 6$ and $i \in \{1,2,3,4,5,6\}$. If s and t are non-negative integers with $6t+i < 6s+i+1 \leq 2(6t+i)$, then $n \in \mathsf{Spec}(6t+i, 6s+i+1)$.*

Lemma 8. *Let $n \equiv 5 \pmod 6$ and $i \in \{1, 4\}$. If s and t are non-negative integers with $6t+i < 6s+i+1 \leq 2(6t+i)$, then $n \in \mathsf{Spec}(6t+i, 6s+i+1)$.*

Let $n \equiv 0, 1, 3$ or $4 \pmod 6$, s and t be non-negative integers, and $i \in \{1,2,3,4,5,6\}$ with $6t+i < 6s+i+2 \leq 2(6t+i)$. Then, by Theorem 4, let $\mathcal{B}_1 \in$

GDD$(X, Y, Z; 6(s-t)+2, 12(s-t)+4)$ and, by Theorem 3, let $\mathcal{B}_2 \in$ BIBD$(X \cup Y \cup Z, 3, 1)$. Thus, $(X, YZ; \mathcal{B})$ forms a GDD$(v = 1+n+n, 3, 3, 6t+i, 6s+i+2)$, where $\mathcal{B} = \mathcal{B}_1 \cup (6(2t-s)+i-2)\mathcal{B}_2$.

Let $n \equiv 2$ or $5 \pmod 6$, s and t be non-negative integers, and $i \in \{2, 5\}$ with $6t+i < 6s+i+2 \leq 2(6t+i)$. Then, by Theorem 4, let $\mathcal{B}_1 \in$ GDD$(X, Y, z; 6(s-t)+2, 12(s-t)+4)$ and, by Theorem 3, let $\mathcal{B}_2 \in$ BIBD$(X \cup Y \cup Z, 3, 3)$. Thus, $(X, Y, Z; \mathcal{B})$ forms a GDD$(v = 1+n+n, 3, 3, 6t+i, 6s+i+2)$, where $\mathcal{B} = \mathcal{B}_1 \cup (\frac{1}{3}(6(2t-s)+i-2))\mathcal{B}_2$. Thus, we have the following results.

Lemma 9. *Let $n \equiv 0, 1, 3$ or $4 \pmod 6$ and $i \in \{1, 2, 3, 4, 5, 6\}$. If s and t are non-negative integers with $6t+i < 6s+i+2 \leq 2(6t+i)$, then $n \in \mathsf{Spec}(6t+i, 6s+i+2)$.*

Lemma 10. *Let $n \equiv 2$ or $5 \pmod 6$ and $i \in \{2, 5\}$. If s and t are non-negative integers with $6t+i < 6s+i+2 \leq 2(6t+i)$, then $n \in \mathsf{Spec}(6t+i, 6s+i+2)$.*

Let $n \equiv 1$ or $3 \pmod 6$, s and t be non-negative integers, and $i \in \{1, 2, 3, 4, 5, 6\}$ with $6t+i < 6s+i+3 \leq 2(6t+i)$. Then, by Theorem 4, let $\mathcal{B}_1 \in$ GDD$(X, Y, Z; 6(s-t)+3, 12(s-t)+6)$ and, by Theorem 3, let $\mathcal{B}_2 \in$ BIBD$(X \cup Y \cup Z, 3, 1)$. Thus, $(X, Y, Z; \mathcal{B})$ forms a GDD$(v = 1+n+n, 3, 3, 6t+i, 6s+i+3)$, where $\mathcal{B} = \mathcal{B}_1 \cup (6(2t-s)+i-3)\mathcal{B}_2$.

Let $n \equiv 5 \pmod 6$, s and t be non-negative integers, and $i \in \{3, 6\}$ with $6t+i < 6s+i+3 \leq 2(6t+i)$. Then, by Theorem 4, let $\mathcal{B}_1 \in$ GDD$(X, Y, Z; 6(s-t)+3, 12(s-t)+6)$ and, by Theorem 3, let $\mathcal{B}_2 \in$ BIBD$(X \cup Y \cup Z, 3, 3)$. Thus, $(X, Y, Z; \mathcal{B})$ forms a GDD$(v = 1+n+n, 3, 3, 6t+i, 6s+i+3)$, where $\mathcal{B} = \mathcal{B}_1 \cup (\frac{1}{3}(6(2t-s)+i-3))\mathcal{B}_2$.

Thus, we have the following results.

Lemma 11. *Let $n \equiv 1$ or $3 \pmod 6$ and $i \in \{1, 2, 3, 4, 5, 6\}$. If s and t are non-negative integers with $6t+i < 6s+i+3 \leq 2(6t+i)$, then $n \in \mathsf{Spec}(6t+i, 6s+i+3)$.*

Lemma 12. *Let $n \equiv 5 \pmod 6$ and $i \in \{3, 6\}$. If s and t are non-negative integers with $6t+i < 6s+i+3 \leq 2(6t+i)$, then $n \in \mathsf{Spec}(6t+i, 6s+i+3)$.*

Let $n \equiv 0, 1, 3$ or $4 \pmod 6$, s and t be non-negative integers, and $i \in \{1, 2, 3, 4, 5, 6\}$ with $6t+i < 6s+i+4 \leq 2(6t+i)$. Then, by Theorem 4, let $\mathcal{B}_1 \in$ GDD$(X, Y, Z; 6(s-t)+4, 12(s-t)+8)$ and, by Theorem 3, let $\mathcal{B}_2 \in$ BIBD$(X \cup Y \cup Z, 3, 1)$. Thus, $(X, Y, Z; \mathcal{B})$ forms a GDD$(v = 1+n+n, 3, 3, 6t+i, 6s+i+4)$, where $\mathcal{B} = \mathcal{B}_1 \cup (6(2t-s)+i-4)\mathcal{B}_2$.

Let $n \equiv 2$ or $5 \pmod 6$, s and t be non-negative integers, and $i \in \{1, 4\}$ with $6t+i < 6s+i+2 \leq 2(6t+i)$. Then, by Theorem 4, let $\mathcal{B}_1 \in$ GDD$(X, Y, Z; 6(s-t)+4, 12(s-t)+8)$ and, by Theorem 3, let $\mathcal{B}_2 \in$ BIBD$(X \cup Y \cup Z, 3, 3)$. Thus, $(X, Y, Z, \mathcal{B})$ forms a GDD$(v = 1+n+n, 3, 3, 6t+i, 6s+i+4)$, where $\mathcal{B} = \mathcal{B}_1 \cup (\frac{1}{3}(6(2t-s)+i-4))\mathcal{B}_2$. Thus, we have the following results.

Lemma 13. *Let $n \equiv 0, 1, 3$ or $4 \pmod 6$ and $i \in \{1, 2, 3, 4, 5, 6\}$. If s and t are non-negative integers with $6t+i < 6s+i+4 \leq 2(6t+i)$, then $n \in \mathsf{Spec}(6t+i, 6s+i+4)$.*

Lemma 14. *Let* $n \equiv 2$ *or* $5 \pmod 6$ *and* $i \in \{1,4\}$. *If* s *and* t *are non-negative integers with* $6t+i < 6s+i+4 \leq 2(6t+i)$, *then* $n \in \mathsf{Spec}(6t+i, 6s+i+4)$.

Let $n \equiv 1$ or $3 \pmod 6$, s and t be non-negative integers, and $i \in \{1,2,3,4,5,6\}$ with $6t+i < 6s+i+5 \leq 2(6t+i)$. Then, by Theorem 4, let $\mathcal{B}_1 \in \text{GDD}(X, Y, Z; 6(s-t)+5, 12(s-t)+10)$ and, by Theorem 3, let $\mathcal{B}_2 \in \text{BIBD}(X \cup Y \cup Z, 3, 1)$. Thus, $(X, Y, Z; \mathcal{B})$ forms a $\text{GDD}(v = 1+n+n, 3, 3, 6t+i, 6s+i+5)$, where $\mathcal{B} = \mathcal{B}_1 \cup (6(2t-s)+i-5)\mathcal{B}_2$.

Let $n \equiv 5 \pmod 6$, s and t be non-negative integers, and $i \in \{2,5\}$ with $6t+i < 6s+i+5 \leq 2(6t+i)$. Then, by Theorem 4, let $\mathcal{B}_1 \in \text{GDD}(X, Y, Z; 6(s-t)+5, 12(s-t)+10)$ and, by Theorem 3, let $\mathcal{B}_2 \in \text{BIBD}(X \cup Y \cup Z, 3, 3)$. Thus, $(X, Y, Z; \mathcal{B})$ forms a $\text{GDD}(v = 1+n+n, 3, 3, 6t+i, 6s+i+5)$, where $\mathcal{B} = \mathcal{B}_1 \cup (\frac{1}{3}(6(2t-s)+i-5))\mathcal{B}_2$. Thus, we have the following results.

Lemma 15. *Let* $n \equiv 1$ *or* $3 \pmod 6$ *and* $i \in \{1,2,3,4,5,6\}$. *If* s *and* t *are non-negative integers with* $6t+i < 6s+i+5 \leq 2(6t+i)$, *then* $n \in \mathsf{Spec}(6t+i, 6s+i+5)$.

Lemma 16. *Let* $n \equiv 5 \pmod 6$ *and* $i \in \{2,5\}$. *If* s *and* t *are non-negative integers with* $6t+i < 6s+i+5 \leq 2(6t+i)$, *then* $n \in \mathsf{Spec}(6t+i, 6s+i+5)$.

Combining results in this section, Theorem 1 and Lemma 1, we obtain the following theorem.

Theorem 5. *Let* λ_1 *and* λ_2 *be positive integers with* $\lambda_1 < \lambda_2$ *and* n *be a positive integer. Then* $n \in \mathsf{Spec}(\lambda_1, \lambda_2)$ *if and only if*

1. $3 \mid [\lambda_1 n(n-1) + \lambda_2 n(n+2)]$,
2. $2 \mid [\lambda_1(n-1) + \lambda_2(n+1)]$, *and*
3. $\lambda_2 \leq 2\lambda_1$.

References

1. Bose, R.C., Shimamoto, T.: Classification and analysis of partially balanced incomplete block designs with two associate classes. J. Amer. Statist. Assoc. 47, 151–184 (1952)
2. Chaiyasena, A., Hurd, S.P., Punnim, N., Sarvate, D.G.: Group divisible designs with two association classes. J. Combin. Math. Combin. Comput. 82(1), 179–198 (2012)
3. Colbourn, C.J., Dinitz, D.H. (eds.): Handbook of Combinatorial Designs, 2nd edn. Chapman and Hall, CRC Press, Boca Raton (2007)
4. Fu, H.L., Rodger, C.A.: Group divisible designs with two associate classes: $n = 2$ or $m = 2$. J. Combin. Theory Ser. A 83(1), 94–117 (1998)
5. Fu, H.L., Rodger, C.A., Sarvate, D.G.: The existence of group divisible designs with first and second associates, having block size 3. Ars Combin. 54, 33–50 (2000)
6. Hurd, S.P., Sarvate, D.G.: Group divisible designs with two association classes and with groups of sizes 1, 1, and n. J. Combin. Math. Combin. Comput. 75, 209–215 (2010)

7. Hurd, S.P., Sarvate, D.G.: Group association designs with two association classes and with two or three groups of size 1. J. Combin. Math. Combin. Comput., 179–198 (2012)
8. Hurd, S.P., Sarvate, D.G.: Group divisible designs with three unequal groups and larger first index. Discrete Math. 311, 1851–1859 (2011)
9. Lindner, C.C., Rodger, C.A.: Design Theory, 2nd edn. CRC Press, Boca Raton (2009)
10. Lapchinda, W., Punnim, N., Pabhapote, N.: GDDs with two associate classes and with three groups of sizes 1, n and n (submitted)
11. Punnim, N., Uiyyasathian, C.: Group divisible designs with two associate classes and $(\lambda_1, \lambda_2) = (1, 2)$. J. Combin. Math. Combin. Comput. 82(1), 117–130 (2012)
12. Pabhapote N., Punnim, N.: Group divisible designs with two associate classes and $\lambda_2 = 1$. Int. J. Math. Math. Sci. 2011, Article ID 148580, 10 pages (2011)

The Number of Diagonal Transformations in Quadrangulations on the Sphere

Naoki Matsumoto[1] and Atsuhiro Nakamoto[2]

[1] Graduate School of Environment and Information Sciences, Yokohama National University, 79-2 Tokiwadai, Hodogaya-ku, Yokohama 240-8501, Japan
matsumoto-naoki-gn@ynu.ac.jp

[2] Department of Mathematics, Yokohama National University, 79-2 Tokiwadai, Hodogaya-ku, Yokohama 240-8501, Japan
nakamoto@ynu.ac.jp

Abstract. A *quadrangulation* is a spherical map of a simple graph such that each face is bounded by a cycle of length four. Since every quadrangulation G is bipartite, G has a unique bipartition $V(G) = B \cup W$, where we call $(|B|, |W|)$ the *bipartition size* of G. In this article, we shall prove that any two quadrangulations G and G' with the same bipartition size can be transformed into each other by at most $10|B| + 16|W| - 64$ diagonal slides.

1 Introduction

A *triangulation* (resp., a *quadrangulation*) is a spherical map of a simple graph such that each face is bounded by a cycle of length 3 (resp., 4). In triangulations, a *diagonal flip* is an operation flipping an edge as shown in Figure 1. If this transformation breaks the simpleness of graphs, then we don't apply it. For diagonal flips of triangulations, the following theorem was proved. For related topics, see a survey [8].

Theorem 1 (Wagner [9]). *Any two triangulations with the same number of vertices can be transformed into each other by diagonal flips.*

In quadrangulations, a *diagonal slide* is an operation sliding an edge as shown in Figure 2, and a *diagonal rotation* is one rotating a path of length 2, where the middle vertex has degree 2 as shown in Figure 3. They are called *diagonal transformations* in quadrangulations. If these transformations break the simpleness of graphs, then we don't apply them.

Since any quadrangulation is bipartite, we always consider a fixed vertex 2-coloring by black and white. For a bipartite graph G (resp., G'), we denote the set of black vertices and that of white vertices by B and W (resp., B' and W'), respectively. Moreover, the *bipartition size* of G means $(|B|, |W|)$. It is easy to see that a diagonal slide preserves the bipartition size, but a diagonal rotation does not. For quadrangulations, Nakamoto [4] proved the following theorems. (In fact, Nakamoto [4] proved similar theorems also for non-spherical surfaces. For related topics, see [3,5].)

J. Akiyama, M. Kano, and T. Sakai (Eds.): TJJCCGG 2012, LNCS 8296, pp. 110–119, 2013.

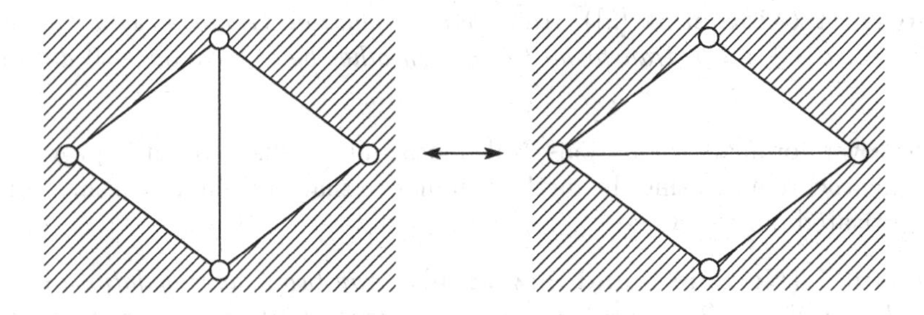

Fig. 1. A diagonal flip

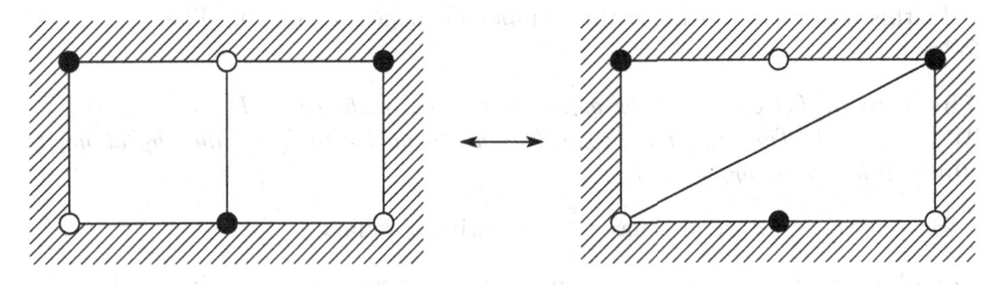

Fig. 2. A diagonal slide

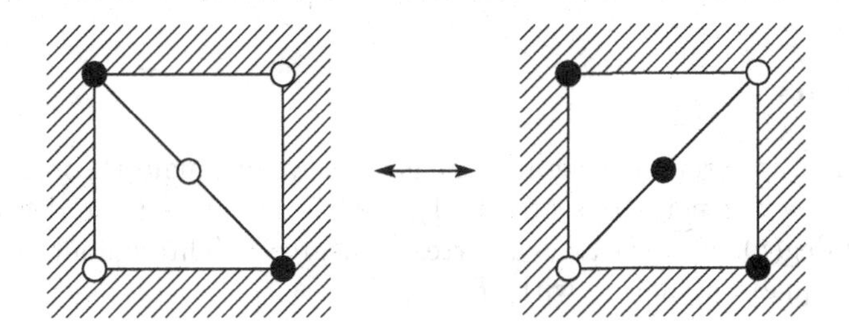

Fig. 3. A diagonal rotation

Theorem 2 (Nakamoto [4]). *Any two quadrangulations with the same number of vertices can be transformed into each other by diagonal transformations.*

Theorem 3 (Nakamoto [4]). *Any two quadrangulations G and G' with the same bipartition size can be transformed into each other only by diagonal slides.*

In this paper, we focus on the number of diagonal transformations in triangulations and quadrangulations. From the proofs of Theorems 1, 2 and 3, we can obtain the fact that any two triangulations (or quadrangulations) can be transformed into each other by $O(n^2)$ transformations, where n is the number of vertices. However, for triangulations, Komuro [1] proved that the number is at most $8n - 48$. Afterward, Mori et al. [2] improved Komuro's result as follows by a clever idea.

Theorem 4 (Mori et al. [2]). *Any two triangulations G and G' with $|V(G)| = |V(G')| = n \geq 6$ can be transformed into each other by at most $6n - 30$ diagonal flips.*

Moreover, for quadrangulations, Nakamoto and Suzuki [6] recently proved the following theorem by using the method similar to that in Komuro's result, which has motivated our study.

Theorem 5 (Nakamoto and Suzuki [6]). *Any two quadrangulations G and G' with $|V(G)| = |V(G')| = n \geq 6$ can be transformed into each other by at most $6n - 32$ diagonal transformations.*

In this paper, we estimate the number of diagonal slides in Theorem 3, as follows.

Theorem 6. *Let G and G' be quadrangulations with $|B| = |B'| = m \geq |W| = |W'| = n \geq 3$. Then G and G' can be transformed into each other by at most $10m + 16n - 64$ diagonal slides.*

By this theorem, we also have the following corollary.

Corollary 1. *Let G and G' be quadrangulations with $k \geq 6$ vertices and $|B| \geq |B'|$. Then there exists a sequence of diagonal transformations of length at most $13k - 64$ from G to G' in which exactly $|B| - |B'|$ diagonal rotations are applied.*

2 Lemmas

In this section, we prepare several lemmas to prove our main theorem. For each vertex v, the set of neighbors of v is denoted by $N(v)$ and the degree of v is denoted by $\deg(v)$. A *k-vertex* is a vertex of degree k. Throughout this section, G denotes a quadrangulation with $|B| \geq |W|$.

Lemma 1. *A 2-vertex in G can be moved to a neighboring face by exactly two diagonal slides.*

Proof. As shown in Figure 4, a 2-vertex can be moved to a neighboring face by exactly two diagonal slides. □

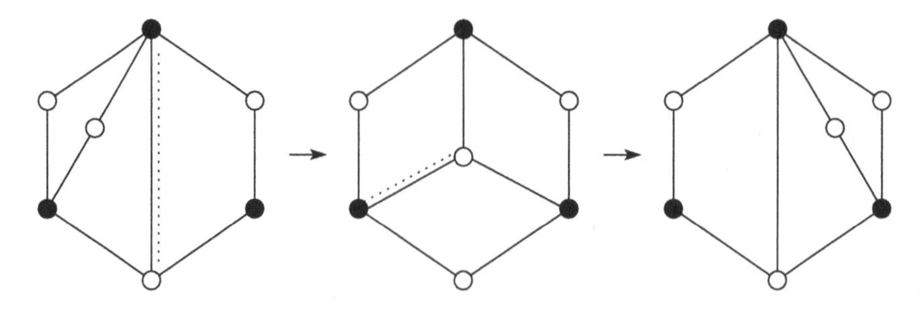

Fig. 4. Moving 2-vertex

Lemma 2. *There exists a vertex $x \in B$ such that* $\deg(x) \leq 3$.

Proof. We suppose that $\deg(x) \geq 4$ for any vertex $x \in B$. Observe $4|F(G)| = 2|E(G)|$, $|V(G)| = |B| + |W|$, and $|E(G)| \geq 4|B|$ by the assumption. By Euler's formula, we have $|E(G)| \leq 2|V(G)| - 4$, and hence,

$$2|B| \leq 2|W| - 4.$$

This inequality contradicts $|B| \geq |W|$. □

Lemma 3. *Let $e = xy \in E(G)$. If the degree of x and y are at least* 3*, then e can be flipped by a diagonal slide.*

Proof. Let $xaby$ and $xcdy$ be two faces sharing the edge xy (note that $a \neq c$ and $b \neq d$ since $\deg(x) \geq 3$ and $\deg(y) \geq 3$). Consider to switch xy to ad by a diagonal slide. If this operation is not applicable, then $ad \in E(G)$. In this case, we can switch xy to cb by a diagonal slide since if $cb \in E(G)$, then G contains a complete bipartite graph $K_{3,3}$ as a subgraph, and so this contradicts G is planar. □

Lemma 4. *At most one diagonal slide yields a* 2*-vertex in B.*

Proof. If G has a 2-vertex $v \in B$, then we are done. Hence we may suppose that there exists $v \in B$ with $\deg(v) = 3$ by Lemma 2. For three vertices $w_1, w_2, w_3 \in N(v)$, if $\deg(w_i) = 2$ for each $i \in \{1, 2, 3\}$, then G has exactly three faces vw_1bw_2, vw_2bw_3 and vw_3bw_1 meeting at $v \in B$, and $B = \{v, b\}$, $W = \{w_1, w_2, w_3\}$. However, this contradicts $|B| \geq |W|$, and hence, we may suppose that $\deg(w_1) \geq 3$. Then, by Lemma 3, we can switch an edge vw_1 to reduce $\deg(v)$ by one. Hence, the lemma holds. □

3 Proof of Theorem 6

In this section, we shall prove Theorem 6. We first prove the following proposition. The quadrangulation shown in Figure 5 is a *standard form* $\mathcal{S}_{m,n}$ which consists of the *equator abcd* (a and c are black and b and d are white), $m - 2$ black 2-vertices in the northern hemisphere which are adjacent to b and d, and $n - 2$ white 2-vertices in the southern hemisphere which are adjacent to a and c.

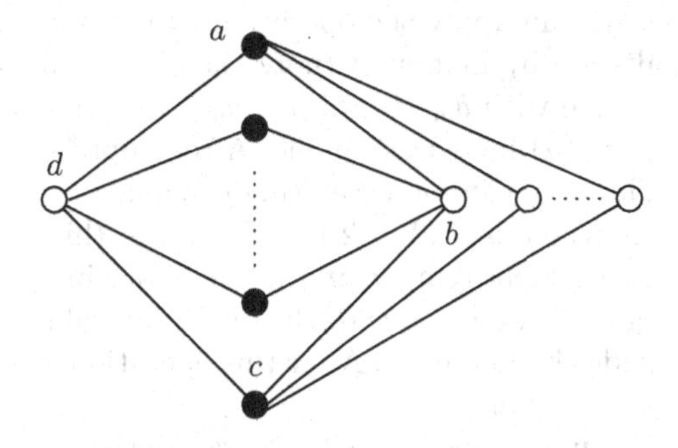

Fig. 5. The standard form $\mathcal{S}_{m,n}$

The following is essential for proving Theorem 6.

Proposition 1. *Let G be a quadrangulation with $|B| = m \geq |W| = n \geq 3$. Then G can be transformed into the standard form by at most $5m + 8n - 32$ diagonal slides.*

Proof. By Lemma 4, we may suppose that there exists $b_1 \in B$ such that $\deg(b_1) = 2$ after we apply at most one diagonal slide. Then, we first regard a face ub_1vb_m as an outer face as shown in Figure 6. Consider the constant $d(u,v) = 2\deg(u) + \deg(v)$. Since $\deg(u) \geq 3$ and $\deg(v) \geq 3$, we have $d(u,v) \geq 9$.

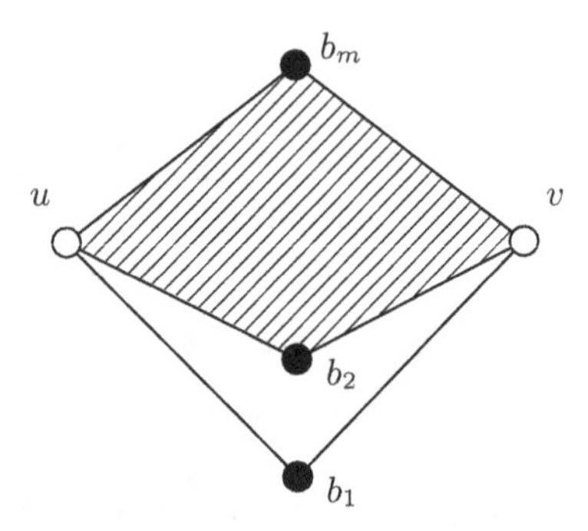

Fig. 6. The first setting

Let ub_1vb_2 be the face sharing the path ub_1v with the outer face ub_1vb_m and let w be a neighbor of b_2 which is next to v with respect to the anti-clockwise rotation around b_2. If $w = u$, then we consider a next vertex $b_3 \in B$ since we have $\deg(b_2) = 2$, where a face ub_2vb_3 shares a path ub_2v with the face ub_1vb_2 (in this case, we regard b_2 and b_3 as b_1 and b_2, respectively). Hence we may suppose $w \neq u$. Let vb_2wx and $wb_2w'y$ be faces sharing the edge b_2w as shown in Figure 7.

The first case is when $\deg(w) \geq 3$ (see Figure 7). Switch b_2w to vy by a diagonal slide. This operation increases $d(u,v)$ by one and decreases $\deg(b_2)$ by one. If this operation is not applicable, then there already exists an edge vy (note that if $b_m = x$, then we can apply the operation). In this case, we can slide b_2v to ux by a diagonal slide by Lemma 3 (note that it also preserves $\deg(v) \geq 3$ since x does not coincide with b_m even if $y = b_m$), and this operation increases $d(u,v)$ by one since $\deg(u)$ increases by one. After applying this operation, we replace x with b_2. Thus, we can increase $d(u,v)$ by one.

The second case is when $\deg(w) = 2$ (see Figure 8; this configuration is obtained from Figure 7 by identifying x and y and replacing them by z). In this case, as shown in Figure 9, we move w to the quadrilateral region ub_1vb_m by exactly four diagonal slides by Lemma 1. After this operation, $d(u,v)$ is un-changed but $\deg(b_2)$ is decreased by one.

Then, since we eventually have $\deg(b_2) = 2$, we continue the above operations for b_3, where ub_2vb_3 is a face sharing a path ub_2v with ub_1vb_2. (In this case, if the

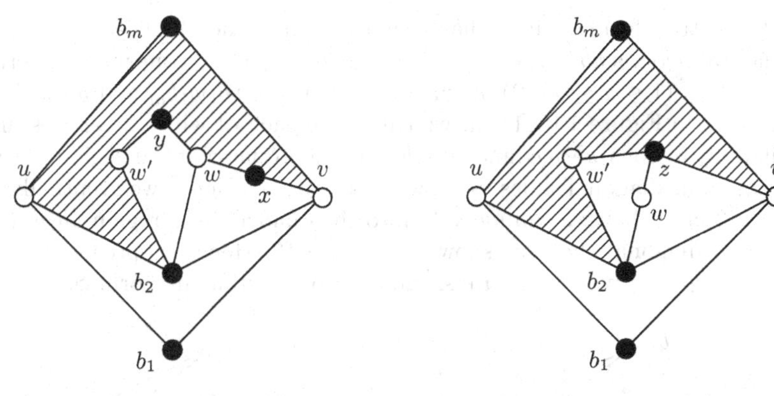

Fig. 7. Case of $\deg(w) \geq 3$

Fig. 8. Case of $\deg(w) = 2$

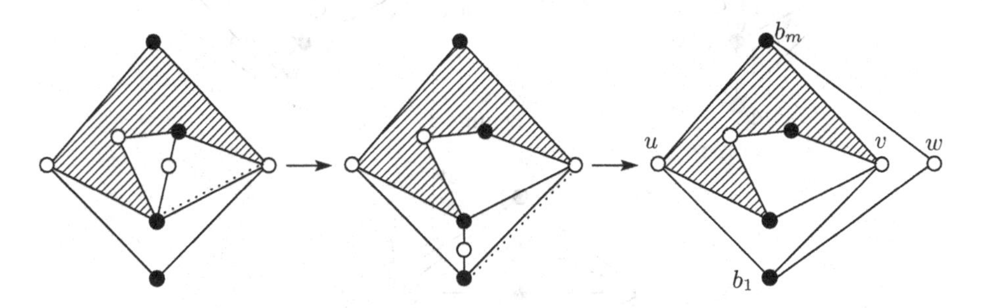

Fig. 9. Moving a white 2-vertex by Lemma 1

above second case appears, then we move a white 2-vertex to the quadrilateral region ub_1vb_2 by exactly four diagonal slides. In general, in the above procedures for b_i, we move a white 2-vertex to the quadrilateral region $ub_{i-2}vb_{i-1}$ by exactly four diagonal slides when the second case happens.) By repeating the above operations, we can transform G into a special form, say H, with $d(u,v) = 3m$ as shown in Figure 10 since u and v are adjacent to all black vertices.

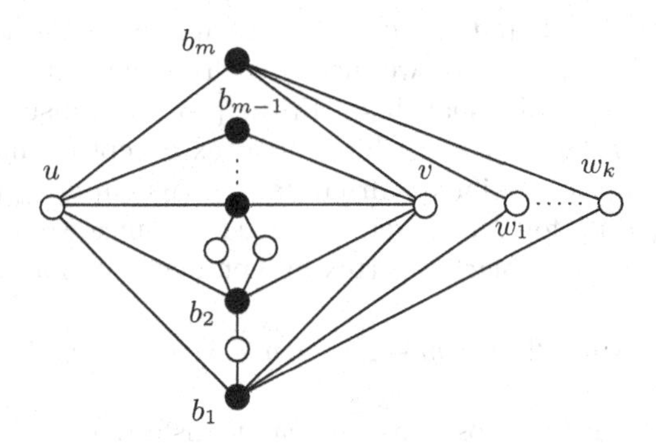

Fig. 10. A quadrangulation H with $d(u,v) = 3m$

Next, we transform H into the standard form. Since $|W| \geq 3$, one of the quadrilateral regions ub_ivb_{i+1}, say ub_mvb_1, includes at least one white 2-vertices, say $w_1, \ldots, w_k$ $(1 \leq k \leq n-2)$, in the interior, where $vb_1w_1b_m$ and $ub_1w_kb_m$ are faces of H (see Figure 10). Then, we can move each of white 2-vertices in two quadrilateral regions ub_1vb_2 and ub_2vb_3 into the quadrilateral region $vb_1w_1b_m$ by at most four diagonal slides as shown in Figure 11. Hence we eventually have $\deg(b_2) = 2$, and then we can move b_2 into the quadrilateral region $ub_1w_kb_m$ by exactly two diagonal slides as shown in Figure 12. Hence, applying the above operations repeatedly, we can transform G into the standard form $\mathcal{S}_{m,n}$.

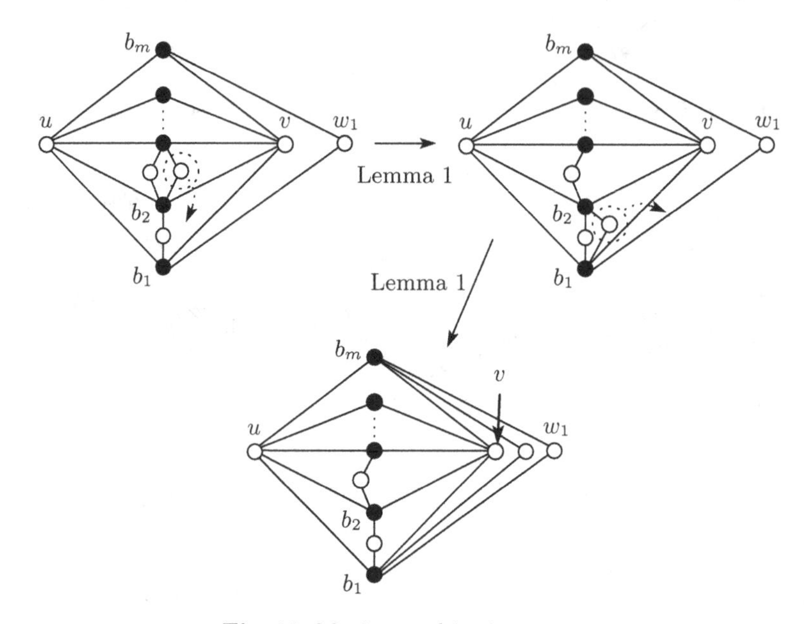

Fig. 11. Moving a white 2-vertex

Finally, we estimate the total number of diagonal slides applied. In the procedures transforming G into H, since one diagonal slide increases $d(u, v)$ by exactly one, and since $d(u, v) \geq 9$ and $d(u, v) = 3m$, we applied at most $3m - 9$ diagonal slides. Moreover, in the process, we moved each of the white 2-vertices except u and v by exactly four diagonal slides. In the process transforming H into the standard form, each black (resp., white) 2-vertex except b_1 and b_m (resp., u, v and w) is moved to the specific quadrilateral regions $ub_1w_1b_m$ (resp., $vb_1w_kb_m$) by at most two (resp., four) diagonal slides. Then, since we apply at most one diagonal slide to obtain a black 2-vertex by Lemma 4, we have

$$(3m - 9) + 4(n - 2) + 2(m - 2) + 4(n - 3) + 1 = 5m + 8n - 32. \square$$

Proof of Theorem 6. By Proposition 1, we can transform G and G' into the same standard form $\mathcal{S}_{m,n}$ by at most $5m+8n-32$ diagonal slides. Therefore, the number of diagonal slides to transform G into G' is at most $10m + 16n - 64$ in total. $\square$

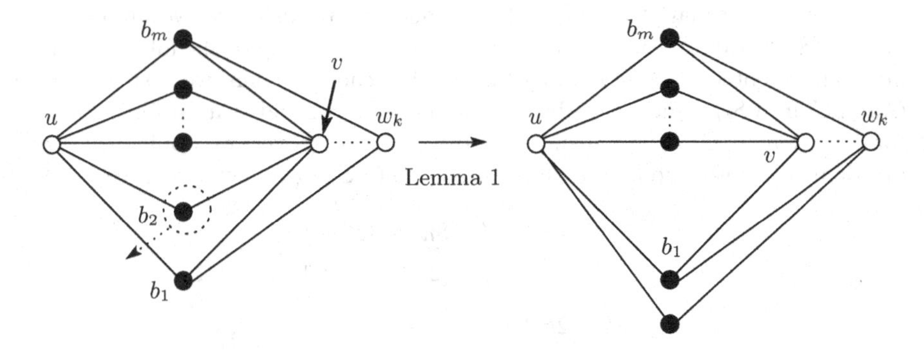

Fig. 12. Moving a black 2-vertex

Finally, we shall prove Corollary 1.

Proof of Corollary 1. Let (m, n) and (m', n') be the bipartition sizes of G and G', respectively. By Proposition 1, G and G' can be transformed into $\mathcal{S}_{m,n}$ and $\mathcal{S}_{m',n'}$ only by diagonal slides, respectively. As shown in Figure 13, we can transform $\mathcal{S}_{k,l}$ into $\mathcal{S}_{k+1,l-1}$ (or $\mathcal{S}_{k-1,l+1}$) by two diagonal slides and one diagonal rotation.

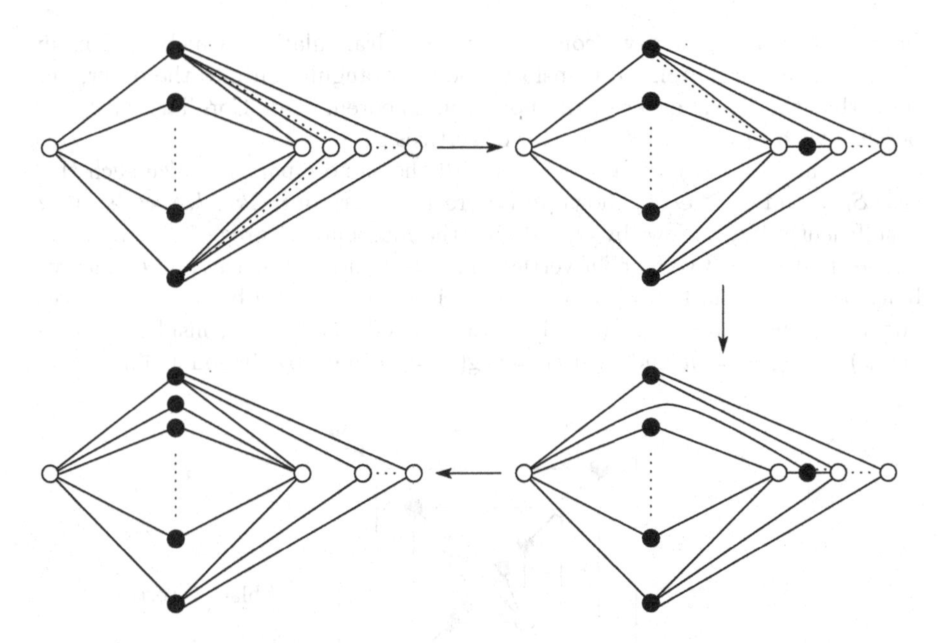

Fig. 13. Transforming a standard form into another standard form

Therefore, since we can transform $\mathcal{S}_{m,n}$ into $\mathcal{S}_{m',n'}$ by applying the operation shown in Figure 13 repeatedly, there exists a sequence of diagonal transformations, where the number of diagonal rotations is exactly $|B| - |B'|$.

Finally, we estimate the length of the sequence. In this proof, we suppose that $m \geq n$. (Since we can easily prove the case $m < n$ similarly to the following proof method, we entrust the remaining case to the reader.) Then, we first transform G and G' into $\mathcal{S}_{m,n}$ and $\mathcal{S}_{m',n'}$ by at most $5m+8n-32$ and at most $5m'+8n'-32$ (or $5n'+8m'-32$) diagonal slides, respectively. Moreover, we apply $3(m-m')$ diagonal transformations to transform $\mathcal{S}_{m,n}$ into $\mathcal{S}_{m',n'}$. Then, we have

$$\begin{aligned}(5m+8n-32)+(5m'+8n'-32)+3(m-m')\\ =(8m+8n)+(2m'+2n')+6n'-64\\ \leq 8k+2k+(6\times\frac{k}{2})-64=13k-64, \text{ or}\\ (5m+8n-32)+(5n'+8m'-32)+3(m-m')\\ =(8m+8n)+(5n'+5m')-64\\ \leq 8k+5k-64=13k-64.\end{aligned}$$

In both cases, since the length of the sequence is at most $13k-64$, we complete the proof. □

4 Examples

In the end of the paper, we construct two quadrangulations which need many diagonal transformations to transform the quadrangulation into the other, and show that the linear order of the bound in Theorem 6 (or Corollary 1) is best possible with respect to the number of vertices.

Let G and G' be quadrangulations with the same bipartition size such that $G \cong \mathcal{S}_{m,n}$ and $G' \cong G_{m,n}$ shown in Figure 14, where $m = 2k + l$, $n = 2k$ and k is sufficiently large. Now, by considering the maximum degree of G and G', we can see that G has two black n-vertices and two white m-vertices and G' has two black 4-vertices and two white $(l+3)$-vertices. Let x and y be black 4-vertices, and let u and v be white $(l+3)$-vertices in G'. Then, we consider to make $\deg(x) = \deg(y) = n$ and $\deg(u) = \deg(v) = m$ only by diagonal slides. Note

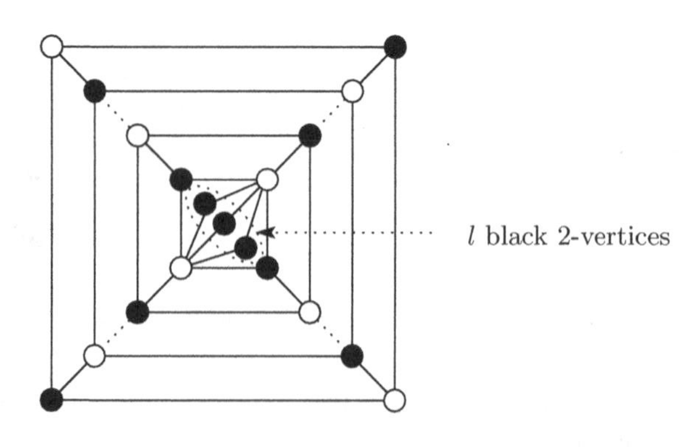

Fig. 14. $G_{m,n}$

that each diagonal slide cannot simultaneously increase the degree of two black (or white) vertices. Moreover, at most four diagonal slides can simultaneously increase the degree of two of x, y, u and v by the simpleness. Thus, we need at least $2(n-4)+2\{m-(l+3)\}-4 = 2m+2n-2l-18$ diagonal slides to transform G into G'. (Note that for any way of transformations, the number of transformations deforming G into G' is at least $2m+2n-2l-18$ since we now choose vertices of the maximum degree.) Therefore, the linear order of the bound in Theorem 6 is best possible.

References

1. Komuro, H.: The diagonal flips of triangulations on the sphere. Yokohama Math. J. 44, 115–122 (1997)
2. Mori, R., Nakamoto, A., Ota, K.: Diagonal flips in Hamiltonian triangulations on the sphere. Graphs Combin. 19, 413–418 (2003)
3. Nakamoto, A.: Diagonal transformations and cycle parities of quadrangulations on surfaces. J. Combin. Theory Ser. B 67, 202–211 (1996)
4. Nakamoto, A.: Diagonal transformations in quadrangulations on surfaces. J. Graph Theory 21, 289–299 (1996)
5. Nakamoto, A.: Quadrangulations on closed surfaces. Interdiscip. Inform. Sci. 7, 77–98 (2001)
6. Nakamoto, A., Suzuki, Y.: Diagonal slides and rotations in quadrangulations on the sphere. Yokohama Math. J. 55, 105–112 (2010)
7. Negami, S.: Diagonal flips in triangulations on surfaces. Discrete Math. 135, 225–232 (1994)
8. Negami, S.: Diagonal flips of triangulations on surfaces, a survey. Yokohama Math. J. 47, 1–40 (1999)
9. Wagner, K.: Bemekungen zum Vierfarbenproblem. J. der Deut. Math. 46, Abt. 1, 26–32 (1936)

Remarks on Schur's Conjecture

Filip Morić and János Pach

Ecole Polytechnique Fédérale de Lausanne
{filip.moric,janos.pach}@epfl.ch

Abstract. Let P be a set of $n > d$ points in $\mathbb{R}^d$ for $d \geq 2$. It was conjectured by Schur that the maximum number of $(d-1)$-dimensional regular simplices of edge length diam(P), whose every vertex belongs to P, is n. We prove this statement under the condition that any two of the simplices share at least $d-2$ vertices. It is left as an open question to decide whether this condition is always satisfied. We also establish upper bounds on the number of all 2- and 3-dimensional simplices induced by a set $P \subset \mathbb{R}^3$ of n points which satisfy the condition that the lengths of their sides belong to the set of k largest distances determined by P.

1 Introduction

The investigation of the distribution of distinct distances induced by a finite set of points in Euclidean space was initiated by Erdős in 1946. It has become a classical topic in discrete and computational geometry, with applications in combinatorial number theory, the theory of geometric algorithms, pattern recognition, etc. A typical problem in the area is Erdős' unit distance problem [2,11]: what is the maximum number of unit distance pairs among n points in $\mathbb{R}^d$?

In the present paper, we concentrate on graphs of diameters. The *diameter graph* $D(P)$ of a finite set of points P in $\mathbb{R}^d$ is the graph whose vertex set is P, and two vertices are connected by an edge if and only if their distance is the diameter of P.

Throughout this paper, d will always denote an integer which is at least 2.

One of the basic properties of graphs of diameters was formulated by Erdős [2]: the maximum number of diameters among n points in the plane is n., Erdős generously attributed the statement to Hopf and Pannwitz [4], who in fact proved a slightly different statement. In 3 dimensions, a similar result was conjectured by Vázsonyi and proved by Grünbaum [5], Heppes [6], and Straszewicz [12]: the maximum number of diameters generated by $n > 3$ points in $\mathbb{R}^3$ is $2n-2$. In higher dimensions, the analogous problem turned out to have a different flavor: Lenz found some simple constructions with a quadratic number of diameters.

In [10], instead of counting the number of edges, Schur, Perles, Martini, and Kupitz initiated the investigation of the number of cliques in a graph of diameters. A *k-clique*, that is, a complete subgraph of k vertices, in the graph of diameters of P corresponds to a regular $(k-1)$-dimensional simplex (or, in short, $(k-1)$*-simplex*) of side length diam(P) generated by P.

J. Akiyama, M. Kano, and T. Sakai (Eds.): TJJCCGG 2012, LNCS 8296, pp. 120–131, 2013.

Theorem A (Schur et al.). *Any finite subset $P \subset \mathbb{R}^d$ contains the vertices of at most one regular d-simplex of edge length* $\mathrm{diam}(P)$.

The main result in [10] is the following.

Theorem B (Schur et al.). *Any set P of n points in $\mathbb{R}^3$ can generate at most n equilateral triangles of side length* $\mathrm{diam}(P)$.

Theorem B can be regarded as another 3-dimensional generalization of the Hopf-Pannwitz result, according to which any set of n points in the plane has at most n diameters. It was conjectured by Z. Schur (see [10]) that this result can be extended to all dimensions d.

Conjecture 1 (Schur). The number of d-cliques in a graph of diameters on n points in $\mathbb{R}^d$ is at most n.

The fact that this bound is tight for any $n > d$ can be shown by the following simple construction given in [10]. Let $p_0, p_1, \ldots, p_d$ be the vertices of a regular d-simplex inscribed in the unit sphere. The edge length of the simplex is $\lambda_d = \sqrt{2(1+1/d)}$. Denote by c the center of the $(d-2)$-simplex $p_0 p_1 \ldots p_{d-2}$. Consider the circle centered at c and passing through p_{d-1} and p_d, and let $p_{d+1}, p_{d+2}, \ldots, p_{n-1}$ be arbitrary points on the short arc between p_{d-1} and p_d of the circle. It is not difficult to see that the set $P = \{p_0, p_1, \ldots, p_{n-1}\}$ has diameter λ_d and determines exactly n regular $(d-1)$-simplices of edge length λ_d. Figure 1 illustrates the case $d = 3$ of this construction.

In a recent manuscript Kupavskii proved Conjecture 1 for $d = 4$.

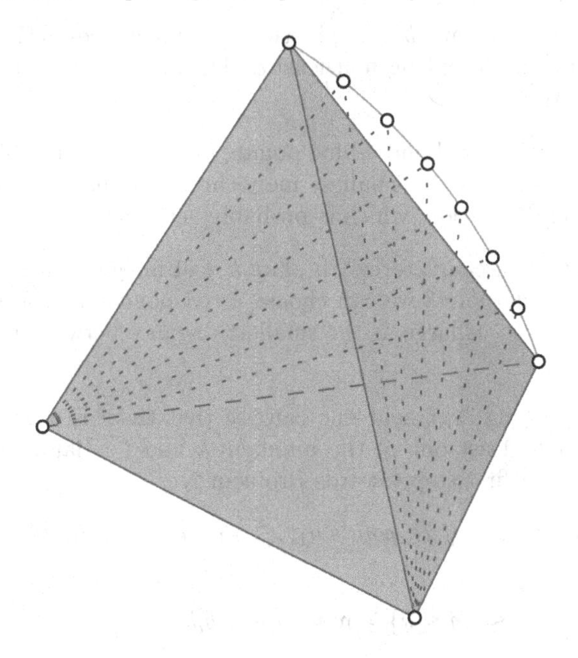

Fig. 1. Construction for $d = 3$

We can prove Schur's conjecture for point sets satisfying a special condition.

Theorem 1. *The number of d-cliques in a graph of diameters on n vertices in* $\mathbb{R}^d$ *is at most n, provided that any two d-cliques share at least* $d-2$ *vertices.*

We do not have any example violating the additional condition and we believe that, in fact, it holds for all graphs of diameters. However, we were unable to prove that it is true in general.

Problem 1. Is it true that any two unit regular $(d-1)$-simplices in $\mathbb{R}^d$ must share at least $d-2$ vertices, provided the diameter of their union is one?

This is vacuously true for $d=2$. For $d=3$ it follows, e.g., from Dolnikov's theorem [1,14] (a direct proof is given in [10]), and it is open for $d \geq 4$. We cannot even verify that two simplices must share at least *one* vertex (for $d \geq 4$), so this step would already be a breakthrough. We propose the following still weaker conjecture.

Conjecture 2. Given two unit regular $(d-1)$-simplices in $\mathbb{R}^d$ with $d \geq 3$, we can choose a vertex u of one simplex and a vertex v of the other one, so that $|uv| \geq 1$.

This is only known to be true for $d=3$. Obviously, a positive answer to Problem 1 would imply Conjecture 2. It seems that regularity of the simplices is not a crucial condition in Conjecture 2, and the following stronger version may be true.

Conjecture 3. Let $a_1 \dots a_d$ and $b_1 \dots b_d$ be two $(d-1)$-simplices in $\mathbb{R}^d$ with $d \geq 3$, such that all their edges have length at least α. Then there exist $i, j \in \{1, \dots, d\}$ such that $|a_i b_j| \geq \alpha$.

In other words, given d red and d blue points, we can find a red-blue distance that is at least as large as the smallest monochromatic distance. We can ask another more general question, which is probably very hard.

Problem 2. For given d, characterize all pairs k, ℓ of integers such that for any set of k red and ℓ blue points we can choose a red point r and a blue point b such that $|rb|$ is at least as large as the smallest distance between two points of the same color.

From an easy packing argument one can see that there is a good choice of r and b, whenever at least one of the numbers k and ℓ is large enough. The following theorem is a first step towards Problem 2.

Theorem 2. *For any set of* $2k$ *points* $a_1, \dots, a_k, b_1, \dots, b_k$ *in* $\mathbb{R}^d$ *the following inequality holds:*

$$\max\{|a_i b_j| \,:\, 1 \leq i, j \leq k\} \geq \min\{|a_i a_j|, |b_i b_j| \,:\, 1 \leq i < j \leq k\},$$

provided that $k \geq c \cdot \sqrt{d} \cdot 2^{\frac{3d}{2}}$ *with a large enough absolute constant c.*

Some generalizations of Theorems A and B to graphs of the k-th largest distances were established in [9]. In this paper we show how these theorems can be extended to *non-regular* triangles in $\mathbb{R}^3$ whose all sides are large (i.e., among the k largest distances). For a given finite set $P \subset \mathbb{R}^3$, we let $d_1 > d_2 > \dots$ be all distinct inter-point distances generated by point pairs in P, so that by d_k we denote the k-th largest distance generated by P.

Theorem 3. *For any $k \in \mathbb{N}$ there is a constant c_k such that the following holds: any set P of n points in $\mathbb{R}^3$ can generate at most $c_k n$ triangles whose all sides have length at least d_k.*

This can be viewed as a 3-dimensional analogue of the well-known observation by Vesztergombi: the number of pairs at distance d_k among n points in the plane is at most $2kn$ (see [15]). The analogous statement for large non-regular $(d-1)$-simplices in $\mathbb{R}^d$ probably holds for $d \geq 4$ as well, but this is open.

The corresponding result for (not necessarily regular) tetrahedra with large edges in $\mathbb{R}^3$ is somewhat weaker in the sense that the bound depends not only on k, but also on the given tetrahedron. We will see in Section 4 that this kind of dependence is necessary.

Theorem 4. *For any tetrahedron T and any k there is a constant $c(T, k)$ such that the following holds: any finite set P of points in $\mathbb{R}^3$ spans at most $c(T, k)$ tetrahedra congruent to T, provided that all edges of T have length at least d_k.*

If Conjecture 3 holds, then Theorem 4 can be generalized to higher dimensions. As for the planar case, it is an easy exercise to show that, for every k, there is a constant c_k such that any finite set of points in the plane spans at most c_k triangles, whose all sides have length at least d_k.

2 Proof of Theorem 1

We start with two lemmas that are borrowed from [13], where they are attributed to [8].

Lemma 1 (Kupitz et al.). *Let a, b, c, d be points on a 2-sphere of radius at least $1/\sqrt{2}$ such that $\operatorname{diam}\{a, b, c, d\} = 1$ and $|ab| = |cd| = 1$. Then the short great circle arcs ab and cd must intersect.*

The maximum number of diameters in a finite set of points on a 2-sphere is the same as in the plane, as long as the radius of the sphere is large enough, compared to the diameter of the set.

Lemma 2 (Kupitz et al.). *Let S be a 2-sphere of radius at least $1/\sqrt{2}$ in $\mathbb{R}^3$. If a set of n points on S has diameter 1, then the diameter occurs at most n times.*

Next, we establish Theorem 1, which says that Schur's conjecture (Conjecture 1) holds, provided that the given graph of diameters satisfies an additional condition: *any two d-cliques share at least $d-2$ vertices.*

Proof of Theorem 1.. Assume without loss of generality that the diameter of our set is equal to 1. We can also assume that every vertex belongs to at least two d-cliques, since otherwise we can proceed by induction. We start with several geometric observations.

Note that the vertices of a d-clique represent d affinely independent points, so their affine hull is $(d-1)$-dimensional, i.e., a hyperplane. Therefore, the affine hull of the d vertices divides the space into two half-spaces.

We will use the expression *angle* uvw and notation $\alpha(u,v,w)$ to refer to the following set of points:

$$\alpha(u,v,w) = \{\mu_1(u-v) + \mu_2(w-v) \,:\, \mu_1, \mu_2 \geq 0\}.$$

Lemma 3. *If two d-cliques* $a_1 \ldots a_{d-2}xy$ *and* $a_1 \ldots a_{d-2}zt$ *share exactly* $d-2$ *vertices, then the open segment* zt *has exactly one common point with* $\mathrm{aff}(a_1,\ldots,a_{d-2},x,y)$*, which lies inside* $\alpha(x,c,y)$*, where* $c = \frac{a_1+\cdots+a_{d-2}}{d-2}$ *is the center of gravity of* $a_1 \ldots a_{d-2}$.

Proof. Since $|a_i x| = |a_i y| = |a_i z| = |a_i t| = 1$ for all $i = 1,\ldots,d-2$, and

$$|cx| = |cy| = |cz| = |ct| = \sqrt{\frac{d-1}{2(d-2)}},$$

we know that the points x, y, z, t lie on a 2-sphere with center c and radius $\geq 1/\sqrt{2}$ (Figures 2(a), 2(b)). Hence, we can apply Lemma 1 to the points x, y, z, t to conclude that the arcs xy and zt intersect at some point p. But then the segment cp is contained in $\alpha(x,c,y)$ and it is intersected by the open segment zt. Therefore, the

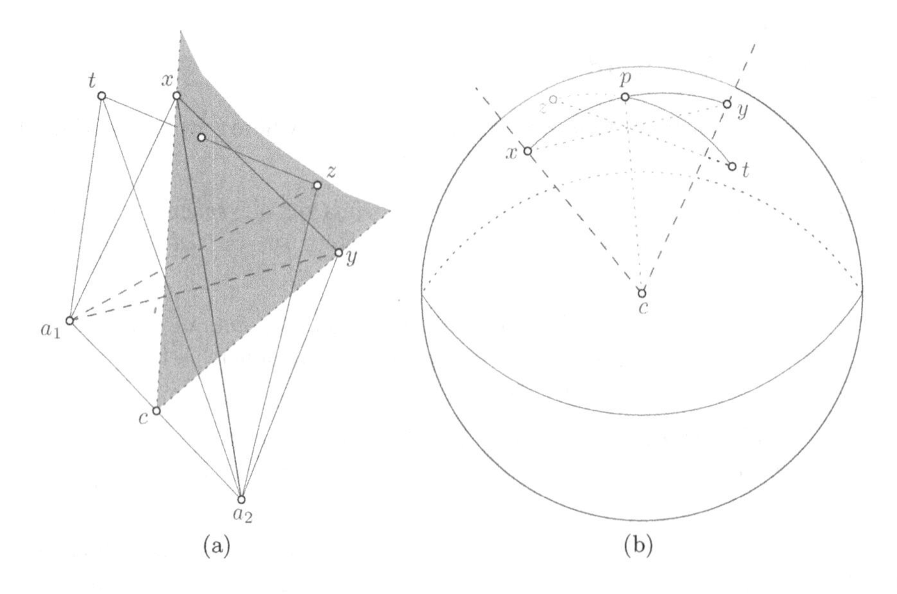

Fig. 2. Proof of Theorem 1, Lemma 3

open segment zt intersects $\text{aff}(a_1, \ldots, a_{d-2}, x, y)$ at a point which lies in $\alpha(x, c, y)$, and in no other point, since otherwise the two d-cliques would lie in the same hyperplane and would necessarily coincide by Theorem A. □

Lemma 4. *There are no three d-cliques that share a $(d-1)$-clique.*

Proof. Suppose the contrary: let $a_1 \ldots a_{d-1}x$, $a_1 \ldots a_{d-1}y$ and $a_1 \ldots a_{d-1}z$ be three d-cliques. Denote by c the center of gravity for $a_1, \ldots, a_{d-1}$. Then the points x, y, z lie on the circle with center c and radius $\sqrt{\frac{d}{2(d-1)}}$, that is orthogonal to $\text{aff}(a_1, \ldots, a_{d-1})$. Since the radius of the circle is at least $1/\sqrt{2}$, we have that $\angle xcy, \angle ycz, \angle zcx \leq \frac{\pi}{2}$. Hence, the points x, y, z lie on a half-circle and we can assume without loss of generality that y is between x and z. Note that $|xy|, |yz| < 1$ and the points x and z lie on different sides of $\text{aff}(a_1, \ldots, a_{d-1}, y)$. According to our initial assumption, there is at least one d-clique C containing y apart from $a_1 \ldots a_{d-1}y$. Since C shares at least $d-2$ points with each of the cliques $a_1 \ldots a_{d-1}x$, $a_1 \ldots a_{d-1}y$ and $a_1 \ldots a_{d-1}z$ and, moreover, C cannot contain x or z, we conclude that C contains exactly $d-2$ of the points $a_1, \ldots, a_{d-1}$. Without loss of generality, let $C = ya_1 \ldots a_{d-2}u$ and let u lie on the same side of $\text{aff}(a_1, \ldots, a_{d-1}, y)$ as x. Now, because of Lemma 3, the open segment $a_{d-1}z$ contains a point from $\alpha(u, c', y)$, where c' is the center of gravity for $a_1, \ldots, a_{d-2}$. However, the whole set $\alpha(u, c', y)$ lies in the closed half-space that contains x, while the open segment $a_{d-1}z$ lies entirely in the open half-space that contains z. This is a contradiction. □

It turns out that the above geometric observations provide enough information so that the proof can be finished more or less combinatorially. We distinguish two cases.

Case 1. *There is a $(d+1)$-clique $a_1 \ldots a_{d+1}$.*

Suppose there is a d-clique C that contains a vertex $x \notin \{a_1, \ldots, a_{d+1}\}$. By the assumption, C shares $d-2$ vertices with the clique $a_1 \ldots a_d$, so we can assume that C contains $a_1, \ldots, a_{d-2}$. But C also shares $d-2$ vertices with the clique $a_2 \ldots a_{d+1}$, so we can also assume that C contains a_{d-1}. Therefore, $C = a_1 \ldots a_{d-1}x$. Thus, we have three d-cliques containing $a_1, \ldots, a_{d-1}$: namely, $a_1 \ldots a_d$, $a_1 \ldots a_{d-1}a_{d+1}$ and C. This is forbidden by Lemma 4. Hence we conclude that all d-cliques must be contained in $a_1 \ldots a_{d+1}$, which gives us at most $d+1$ cliques, so in this case the statement is proven, since $n \geq d+1$.

Case 2. *There is no $(d+1)$-clique.*

We have two subcases.

Subcase 2.1 *There are two d-cliques that share $d-1$ vertices.*

Let the cliques be $a_1 \ldots a_{d-1}x$ and $a_1 \ldots a_{d-1}y$. Observe that $|xy| < 1$, since we assume there is no $(d+1)$-clique. If there are no more d-cliques except for those generated by $a_1, \ldots, a_{d-1}, x, y$, we are done. So we can suppose that there are some more d-cliques. Any new d-clique shares $d-2$ points both with $a_1 \ldots a_{d-1}x$ and

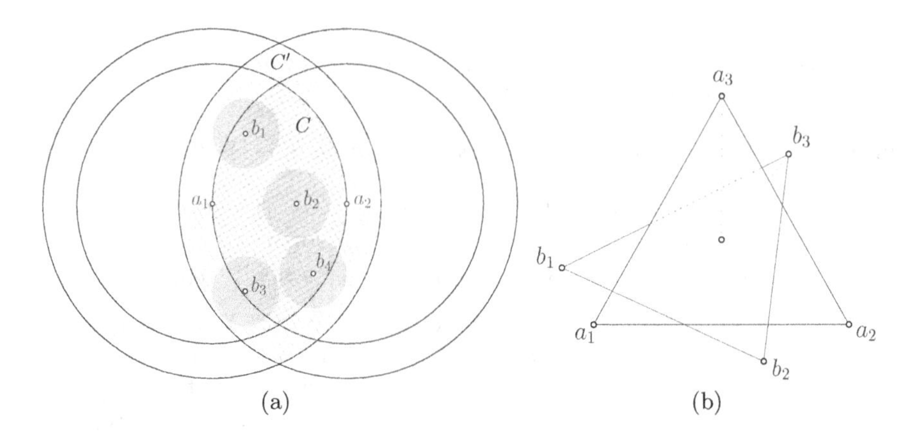

Fig. 3. (a) Proof of Theorem 2; (b) construction for $d = 4$: two equilateral triangles in two orthogonal planes with a common center at the origin

with $a_1 \dots a_{d-1}y$. Hence, any new clique contains *exactly* $d - 2$ of the vertices $a_1, \dots, a_{d-1}$. We say that a d-clique is of *type* k if it contains all the vertices $a_1, \dots, a_{d-1}$ except for a_k. Now we will again branch out into different cases.

First, let us see what happens if all d-cliques have the same type, e.g., they all contain the points $a_1, \dots, a_{d-2}$. The remaining two vertices of any d-clique must lie on the 2-sphere with center $\frac{a_1+\dots+a_{d-2}}{d-2}$ and radius $\sqrt{\frac{d-1}{2(d-2)}} > \frac{1}{\sqrt{2}}$. Thus, the number of d-cliques is no more than the number of unit-diameters among $n - (d - 2)$ points on a 2-sphere of radius $> 1/\sqrt{2}$, which is at most $n-(d-2)$, by Lemma 2.

Therefore, we can assume that there are at least two d-cliques of different types. Any two cliques of different types share exactly $d - 3$ vertices among $a_1, \dots, a_{d-1}$, so they must share at least one more vertex. Again, we consider different cases.

Suppose there are two d-cliques of different types that share a vertex v outside of $\{a_1, \dots, a_{d-1}, x, y\}$. Let the cliques be $a_1 \dots a_{d-2}uv$ and $a_2 \dots a_{d-1}vw$. Clearly, $a_1 \dots a_{d-1}v$ is also a d-clique, so we have three d-cliques sharing $d - 1$ points $a_1, \dots, a_{d-1}$, which is impossible, according to Lemma 4.

The second possibility that remains is that any two cliques of different types contain x or y. This means that either all cliques (apart from the initial two) contain x or all of them contain y. Without loss of generality, let all new cliques contain x. Notice that there can be at most one clique of each type, for if C_1 and C_2 were d-cliques of the same type, say, type 1, there would be three d-cliques sharing $d - 1$ points $x, a_2, \dots, a_{d-1}$, contrary to Lemma 4. Consequently, in this case we have at most $d+1$ cliques, and the total number of vertices is at least $d + 2$.

Subcase 2.2 *Any two d-cliques share at most $d-2$ vertices.*

Let $a_1 \dots a_{d-2}xy$ and $a_1 \dots a_{d-2}zt$ be two d-cliques. None of the points x and y forms a diameter with any of the points z and t, since it would produce two d-cliques that share $d-1$ vertices. If all other cliques contain $a_1, \dots, a_{d-2}$, we are done as above, so without loss of generality suppose that there is a d-clique $a_1 \dots a_{d-3}xuv$. Clearly, u, v are new points, i.e., different from $a_1, \dots, a_{d-2}, x, y, z, t$. But now $a_1 \dots a_{d-3}xuv$ and $a_1 \dots a_{d-2}zt$ have only $d-3$ points in common, contradicting the assumption.

We have proved that n is an upper bound for the number of d-cliques. A construction from [10] showing that this bound can be achieved is given in Introduction. This completes the proof of Theorem 1. □

Remark. If the statement from Problem 1 is true, then Theorem 1 would confirm Schur's conjecture. The following weaker statement might be easier to prove: *There is a constant $K(d)$ such that among any $K(d)$ cliques in a graph of diameters, there are two cliques sharing a vertex.* If true, this would give a bound of the form $k(d) \cdot n$ for Schur's conjecture. However, it appears that even this weaker form requires a new insight.

It is natural to extend Problem 1 to cliques that might have fewer than d vertices. In particular, is it true that a d-clique and a $(d-1)$-clique in a graph of diameters in $\mathbb{R}^d$ must share a vertex? For $d = 2$ and $d = 3$, this is clearly false. It is also false in $\mathbb{R}^4$, as shown by the following construction (for $k = 2$).

Proposition 1. *For every $k \geq 2$, there exist a unit regular $(2k-1)$-simplex and a unit regular k-simplex in $\mathbb{R}^{2k}$ that do not share a vertex, while the diameter of their union is 1.*

Proof. Consider a unit regular $(2k-1)$-simplex $\Delta = a_1 \dots a_{2k}$ in $\mathbb{R}^{2k}$ and let $u_1, \dots, u_k$ be the midpoints of the edges $a_1a_2, a_3a_4, \dots, a_{2k-1}a_{2k}$, respectively. Let the origin $o = (0, \dots, 0)$ be the center of the simplex Δ and let the simplex lie in the hyperplane $x_{2k} = 0$. For every $n \geq 1$, denote by r_n the circumradius of a unit regular n-simplex. We have that $r_n = \sqrt{\frac{n}{2n+2}}$. Denote by $v_1, \dots, v_k$ the points such that $|ov_i| = r_{k-1}$ and u_i lies on the segment ov_i for $i = 1, 2, \dots, k$. Then $v_1v_2 \dots v_k$ is a unit regular $(k-1)$-simplex with center o. Translate the points $v_1, \dots, v_k$ by the vector $(0, \dots, 0, \sqrt{\frac{3-2\sqrt{2}}{4k}})$ to get points $w_1, \dots, w_k$, and let $w_{k+1} = (0, \dots, 0, \sqrt{\frac{3-2\sqrt{2}}{4k}} - \sqrt{\frac{k+1}{2k}})$. Now it is not difficult to verify that $\tilde{\Delta} = w_1 \dots w_k w_{k+1}$ is a unit regular k-simplex and that the pair of simplices Δ and $\tilde{\Delta}$ satisfies the needed conditions (we omit the straightforward calculation). □

The question whether a d-clique and a $(d-1)$-clique in a graph of diameters in $\mathbb{R}^d$ must share a vertex remains open for $d \geq 5$.

3 Proof of Theorem 2

Proof of Theorem 2. Suppose the contrary, i.e., that the maximum is strictly smaller than the minimum, while $k \geq c \cdot \sqrt{d} \cdot 2^{\frac{3d}{2}}$ for a large enough c. Without loss of generality, we assume that

$$\min\{|a_i a_j|, |b_i b_j| \,:\, 1 \leq i < j \leq k\} = 1$$

and $|a_1 a_2| = 1$. Denote by C the intersection of two balls with centers a_1 and a_2 and radius 1 (Figure 3(a)). Then C contains all the points $b_1, \ldots, b_k$. Since $|b_i b_j| \geq 1$, the balls centered at $b_1, \ldots, b_k$ with radii $\frac{1}{2}$ do not overlap. Moreover, all these balls are contained in C', which is the intersection of the balls with centers a_1 and a_2 and radius $\frac{3}{2}$. Let us estimate the volume of C'. Using the fact that the volume of a spherical cap of height h is

$$\frac{\pi^{\frac{d-1}{2}} r^d}{\Gamma\left(\frac{d+1}{2}\right)} \int_0^{\arccos \frac{r-h}{h}} \sin^d(t)\, dt\,,$$

where r is the radius of the sphere, we get

$$\mathrm{Vol}(C') = 2 \cdot \frac{\pi^{\frac{d-1}{2}} (3/2)^d}{\Gamma\left(\frac{d+1}{2}\right)} \int_0^{\arccos \frac{1}{3}} \sin^d(t)\, dt$$

$$\leq 2 \cdot \frac{\pi^{\frac{d-1}{2}} (3/2)^d}{\Gamma\left(\frac{d+1}{2}\right)} \cdot \left(\frac{2\sqrt{2}}{3}\right)^d \cdot \arccos \frac{1}{3} = O\left(\frac{(2\pi)^{\frac{d}{2}}}{\Gamma\left(\frac{d+1}{2}\right)}\right).$$

But C' contains k non-overlapping balls of radius $\frac{1}{2}$, and, therefore,

$$k \cdot \frac{\pi^{\frac{d}{2}} 2^{-d}}{\Gamma\left(1 + \frac{d}{2}\right)} \leq O\left(\frac{(2\pi)^{\frac{d}{2}}}{\Gamma\left(\frac{d+1}{2}\right)}\right).$$

Finally, taking into account the asymptotics $\Gamma(x) \sim x^{x-\frac{1}{2}} e^{-x} \sqrt{2\pi}$, we obtain $k = O(\sqrt{d} \cdot 2^{3d/2})$, with a contradiction, as long as c is large enough. □

Remark. On the other hand, we know that Theorem 2 does not hold with $k \leq \lceil \frac{d+1}{2} \rceil$. To see this, consider the following construction. Let $a_1 \ldots a_k$ be a regular $(k-1)$-dimensional simplex inscribed in the sphere

$$\{(x_1, \ldots, x_d) \,:\, x_1^2 + \cdots + x_{k-1}^2 = 1,\, x_k = \cdots = x_d = 0\}$$

and let $b_1 \ldots b_k$ be a regular $(k-1)$-dimensional simplex inscribed in the sphere

$$\{(x_1, \ldots, x_d) \,:\, x_k^2 + \cdots + x_{2k-2}^2 = 1,\, x_1 = \cdots = x_{k-1} = 0\}.$$

Then $|a_i a_j| = |b_i b_j| = \sqrt{\frac{2k}{k-1}}$ for all $i \neq j$, while $|a_i b_j| = \sqrt{2}$ (Figure 3(b)).

Thus, the smallest $k(d)$ for which Theorem 2 holds is somewhere between $d/2$ and $c\sqrt{d} \cdot 2^{\frac{3d}{2}}$. The gap is obviously quite large, and Conjecture 3 suggests the answer should be closer to the lower bound.

4 Proofs of Theorems 3 and 4

The proofs of Theorem 3 and Theorem 4 are both analogous to the proofs of the corresponding statements for regular simplices given in [9], with the only new ingredient being the next lemma.

Lemma 5. *Let $a_1a_2a_3$ and $b_1b_2b_3$ be two triangles in $\mathbb{R}^3$ such that all their sides have length at least α. Then there exist $i, j \in \{1,2,3\}$ such that $|a_ib_j| \geq \alpha$.*

Proof. Suppose the contrary, i.e., that the two triangles are placed so that $|a_ib_j| < \alpha$ for all i and j. Without loss of generality, let $a_1a_2a_3$ lie in the plane $x_3 = 0$. By the pigeon hole there are two vertices of $b_1b_2b_3$ that lie on the same side of $x_3 = 0$. Without loss of generality, let b_1 and b_2 lie in the half-space $x_3 \geq 0$ and let $b_1 = (0,0,p)$ and $b_2 = (0,q,r)$, where p,q,r are non-negative and $r \geq p$ (Figure 4). Translate the points b_1 and b_2 by the vector $(0,0,-p)$ to get new points $c_1 = (0,0,0)$ and $c_2 = (0,q,r-p)$. Note that $|c_1c_2| = |b_1b_2| \geq \alpha$ and $|c_ia_j| \leq |b_ia_j| < \alpha$ for all $i \in \{1,2\}$, $j \in \{1,2,3\}$. It follows that the points a_1, a_2, a_3 must have non-negative second coordinates. Now we rotate the point c_2 around c_1 in the plane $x_1 = 0$ until it hits the plane $x_3 = 0$. Thus, we replace c_2 by $c_2' = (0,s,0)$, where $s = \sqrt{q^2 + (r-p)^2}$. Again, $|c_1c_2'| = |c_1c_2| \geq \alpha$ and the distances between c_2' and a_j for $j \in \{1,2,3\}$ are all smaller than α. Indeed, letting $a_j = (t,u,0)$, we have

$$|c_2'a_j| = \sqrt{t^2 + (u-s)^2} \leq \sqrt{t^2 + (q-u)^2 + (r-p)^2} = |c_2a_j| < \alpha\,,$$

where we used that $u \geq 0$ and $q \leq s$.

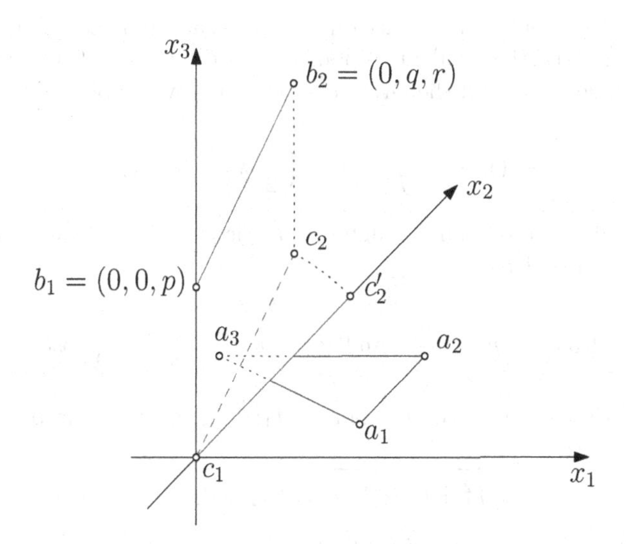

Fig. 4. Proof of Lemma 5

The points c_1, c_2', a_1, a_2, a_3 lie in the same plane and segment c_1c_2' can intersect at most two sides of triangle $a_1a_2a_3$ at their interior points. So, without loss of generality, assume that c_1c_2' does not intersect a_1a_2 at an interior point. Then either c_1, c_2', a_1, a_2 are in convex position or an extension of one of the segments c_1c_2' and a_1a_2 intersects the other one. In either case one can easily show that one of the segments $c_1a_1, c_1a_2, c_2'a_1, c_2'a_2$ has length at least $\min\{|c_1c_2'|, |a_1a_2|\} \geq \alpha$. Contradiction. □

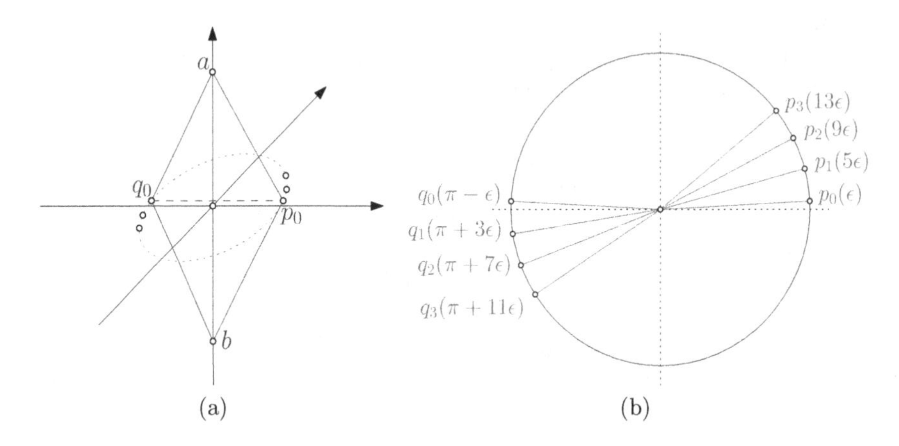

Fig. 5. (a) Construction with many congruent large non-regular simplices; (b) points in the plane $x_3 = 0$

Remark. Note that some dependence on T is necessary in Theorem 4, as shown by this simple construction. Take two points $a = (0, 0, 1)$, $b = (0, 0, -1)$, and $2n$ points in the plane $x_3 = 0$ on the circle $x_1^2 + x_2^2 = 1/4$ with polar coordinates as follows:

$$p_i = \left(\frac{1}{2}, (1+4i)\epsilon\right), \quad q_i = \left(\frac{1}{2}, \pi + (4i-1)\epsilon\right),$$

for $i = 0, 1, \ldots, n-1$ and small enough $\epsilon > 0$ (Figures 5(a),5(b)). In this set of $2n + 2$ points we have that

$$d_1 = |ab| = 2, d_2 = |ap_i| = \frac{\sqrt{5}}{2} \text{ and } d_3 = |p_iq_i| = \sqrt{\frac{1}{2} + \frac{1}{2}\cos(2\epsilon)} < 1\,.$$

Recall that the distance between the points (r_1, θ_1) and (r_2, θ_2) in polar coordinates is equal

$$\sqrt{r_1^2 + r_2^2 - 2r_1r_2\cos(\theta_1 - \theta_2)}.$$

Also, we can check that for all i, j we have

$$|p_iq_j| = \sqrt{\frac{1}{2} + \frac{1}{2}\cos((4(j-i)-2)\epsilon)} \leq \sqrt{\frac{1}{2} + \frac{1}{2}\cos(2\epsilon)} = d_3\,,$$

since $|4(j-i)-2| \geq 2$. It remains to notice that the chosen points span $2n-1$ tetrahedra with edge lengths $d_1, d_2, d_2, d_2, d_2, d_3$. Those are the tetrahedra abp_iq_j for all $i, j \in \{0, 1, \ldots, n-1\}$ such that $j-i \in \{0, 1\}$. Thus, for $k=3$ we can have an arbitrarily large number of tetrahedra whose all edges have lengths at least d_k.

Acknowledgements. The authors gratefully acknowledge support from the Hungarian Science Foundation EuroGIGA Grant OTKA NN 102029, from the Swiss National Science Foundation Grants 200020-144531 and 20021-137574, and from the NSF Grant CCF-08-30272.

References

1. Dolnikov, V.L.: Some properties of graphs of diameters. Discrete Comput. Geom. 24, 293–299 (2000)
2. Erdős, P.: On sets of distances of n points. Amer. Math. Monthly 53, 248–250 (1946)
3. Erdős, P., Pach, J.: Variations on the theme of repeated distances. Combinatorica 10, 261–269 (1990)
4. Hopf, H., Pannwitz, E.: Aufgabe Nr. 167. Jahresbericht Deutsch. Math.-Verein 43, 114 (1934)
5. Grünbaum, B.: A proof of Vázsonyi's conjecture. Bull. Res. Council Israel, Sect. A 6, 77–78 (1956)
6. Heppes, A.: Beweis einer Vermutung von A. Vázsonyi. Acta Math. Acad. Sci. Hungar. 7, 463–466 (1956)
7. Kupavskii, A.: Diameter graphs in $\mathbb{R}^4$ (manuscript, 2013), `http://arxiv.org/abs/1306.3910`
8. Kupitz, Y.S., Martini, H., Wegner, B.: Diameter graphs and full equi-intersectors in classical geometries. In: IV International Conference in "Stoch. Geo., Conv. Bodies, Emp. Meas. & Apps. to Eng. Sci.", vol. II, Rend. Circ. Mat. Palermo (2) Suppl. No. 70, part II, pp. 65–74 (2002)
9. Morić, F., Pach, J.: Large simplices determined by finite point sets, Contr. to Alg. and Geom. 54, 45–57 (2013), for a slightly updated version `http://www.math.nyu.edu/~pach/publications/sim_short_final.pdf`
10. Schur, Z., Perles, M.A., Martini, H., Kupitz, Y.S.: On the number of maximal regular simplices determined by n points in $\mathbb{R}^d$. In: Aronov et al. (eds.) Discrete and Computational Geometry, The Goodman-Pollack Festschrift, pp. 767–787. Springer (2003)
11. Spencer, J., Szemerédi, E., Trotter, W.T.: Unit distances in the Euclidean plane. In: Bollobás, B. (ed.) Graph Theory and Combinatorics, pp. 293–303. Academic Press, London (1984)
12. Straszewicz, S.: Sur un problème géométrique de P. Erdős. Bull. Acad. Pol. Sci., Cl. III 5, 39–40 (1957)
13. Swanepoel, K.J.: Unit distances and diameters in Euclidean spaces. Discrete Comput. Geom. 41, 1–27 (2009)
14. Swanepoel, K.J.: A new proof of Vázsonyi's conjecture. J. Combinat. Th., Series A 115, 888–892 (2008)
15. Vesztergombi, K.: On large distances in planar sets. Discrete Math. 67, 191–198 (1987)

Greedy Approximation Algorithms for Generalized Maximum Flow Problem towards Relation Extraction in Information Networks

Yusuke Nojima, Yasuhito Asano, and Masatoshi Yoshikawa

Department of Social Informatics, Graduate School of Informatics, Kyoto University
nojima@db.soc.i.kyoto-u.ac.jp, {asano,yoshikawa}@i.kyoto-u.ac.jp

Abstract. Generalized maximum flow problem is a generalization of the traditional maximum flow problem, where each edge e has gain factor $\gamma(e)$. When $f(e)$ units of flow enter edge $e = (u, v)$ at u, then $\gamma(e)f(e)$ units of flow arrive at v. Since relation extraction, which is an important application of the problem, uses large networks such as Wikipedia and DBLP, the computation time to solve the problem is important. However, conventional algorithms for the problem are expensive and do not scale to large graph. Therefore, we propose approximation algorithms based on greedy augmentation and a heuristic initial flow calculation. The experimental result shows that our algorithms are two orders of magnitude faster than a conventional algorithm.

1 Introduction

Generalized maximum flow problem is a generalization of the traditional maximum flow problem. In the generalized maximum flow problem, each edge e has a real number called gain factor $\gamma(e)$. When $f(e)$ units of flow enter the edge $e = (u, v)$ at u, then $\gamma(e)f(e)$ units of flow arrive at v. The objective is to maximize the flow entering a distinguished vertex called the sink.

One of the important applications of the generalized maximum flow problem is relation extraction on information networks. An information network is a network in which each edge represents an explicit relation between two objects. For example, if a web page p_1 has a hyper link to a web page p_2, p_1 is considered to have an explicit relation to p_2. Given an information network and two objects, the relation extraction problem is to compute the strength of the implicit relation between the two objects. An implicit relation is represented by a subgraph of the information network including the two objects. For example, if a web page p_1 has a hyper link to a web page p_2, and p_2 has a hyper link to a web page p_3, then p_1 and p_3 are considered to have an implicit relation through p_2.

By our observation, the following characteristics of implicit relations can be assumed:

- If there are many disjoint paths between the given two objects, the implicit relation between the two objects is strong. In Figure 1, the implicit relation

J. Akiyama, M. Kano, and T. Sakai (Eds.): TJJCCGG 2012, LNCS 8296, pp. 132–142, 2013.

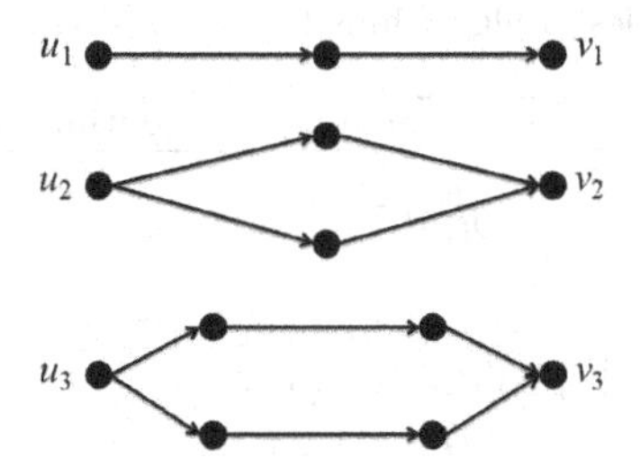

Fig. 1. The examples of implicit relations

between u_2-v_2 is stronger than the implicit relation between u_1-v_1 because u_2-v_2 has two distinct paths and u_1-v_1 has only one distinct path.
- If paths connecting the given two objects are short, the implicit relation between the two objects is strong. In Figure 1, the implicit relation between u_2-v_2 is stronger than the implicit relation between u_3-v_3 because the distance between u_2-v_2 is two and the distance between u_3-v_3 is three.

The relation extraction problem can be modeled by the generalized maximum flow problem [1]. The generalized maximum flow problem can reflect the connectivity between the source and the sink because the generalized maximum flow problem inherits the characteristic of the maximum flow problem. The generalized maximum flow problem can also reflect the distance between the source and the sink because a generalized flow decays in accordance with the length of the path.

Besides the generalized maximum flow model, many other models are proposed to extract implicit relations from information networks. Hitting Time [2,3] and Truncated Hitting Time [3] are similarity measures based on the average lengths of all paths between the given two objects. Co-citation [4] is a similarity measure based on the number of objects cited by both of the given two objects. PFIBF [5] is a relation measure based on cohesion defined relying on the number of paths between the given two objects.

An advantage of the generalized maximum flow model over these models is that the generalized maximum flow model can simultaneously reflect both assumption described above. Zhang, et al. [1] conducted an experiment and concluded that a fairly reasonable implicit relations are obtained by the generalized maximum flow model compared to other models.

Because large networks, such as Wikipedia and DBLP, are typically used in the relation extraction, the computation time of solving the problem is important. Conventional algorithms to solve generalized maximum flow problems are expensive and do not scale to large networks. The best exact algorithm currently known is Goldfarb, Jin and Orlin's $O(m^2(m + n \log n) \log B)$ time algorithm [6], and the best ε-optimum approximation scheme is Fleischer and Wayne's $O(m(m + n \log m)\varepsilon^{-2} \log n)$ time algorithm [7], where n and m are the number of vertices and the number of edges, respectively, and B is the largest integer required for representing capacities and gain factors of the network.

Table 1. Time complexities of algorithms for generalized maximum flow problem

	ε-optimum flow / optimum flow
Kapoor-Vaidya [8]	-
	$O(m^{1.5}n^{2.5}\log B)$
Vaidya [9]	-
	$O(m^{1.5}n^{2}\log B)$
Goldberg-Plotkin-Tardos [10]	$\tilde{O}(mn^2 \log B \log(1/\varepsilon))$
	$\tilde{O}(m^2n^2 \log B)$
Cohen-Megiddo [11]	$\tilde{O}(m^2n^2 \log(1/\varepsilon))$
	$\tilde{O}(m^3n^2 \log B)$
Radzik [12]	$\tilde{O}(m^2n + \min\{m^2n, m(m + n\log\log B)\}\log\frac{1}{\varepsilon})$
	$\tilde{O}(m^2(m + n\log\log B)\log B)$
Tardos-Wayne [13]	$\tilde{O}(m(m + n\log\log B)\log(1/\varepsilon) + mn^2\log B)$
	$\tilde{O}(m^2(m + n\log\log B)\log B)$
Fleischer-Wayne [7]	$O(m(m + n\log m)\varepsilon^{-2}\log n)$
	$\tilde{O}(m^2(m + n\log\log B)\log B)$
Goldfarb-Jin-Orlin [6]	-
	$O(m^2(m + n\log n)\log B)$

We propose approximation algorithms to calculate the generalized maximum flow based on a greedy method. Our algorithm augments a flow along the highest-fatness residual paths. The highest-fatness residual path is a path along which largest amount of flow can reach to the sink. We also discuss the heuristic algorithm for initial flow calculation. Although our algorithms have no accuracy assurances, the experimental result showed that our algorithms accomplish comparable accuracy to Fleischer and Wayne's algorithm with $\varepsilon = 0.2$, and run two orders of magnitude faster than the conventional algorithm.

2 Related Work

There are many studies on the generalized maximum flow problem. We discuss various approaches for solving generalized maximum flow problem.

Because the generalized maximum flow problem is a special case of the linear programming, there are approaches that utilize the linear programming methods. Vaidya [9] proposed the $O(m^{1.5}n^2 \log B)$ time algorithm that utilize the Karmarkar's interior point method, together with speeding up techniques developed by Kapoor and Vaidya [8]. This algorithm can solve a more general version of the generalized maximum flow problem than the one considered in this paper. The time complexity of this algorithm, however, remains same if the input graph is constrained to the generalized maximum flow problem considered in this paper.

Combinatorial approach is another approach to the generalized maximum flow problem. In combinatorial approach, we focus on the combinatorial structure of the given network, and utilize the network algorithms such as shortest path algorithms, maximum flow algorithms, and minimum cost flow algorithms.

Onaga [14] showed that a flow is optimum if and only if neither augmenting paths nor generalized augmenting paths (augmenting paths from flow generating cycles) exist.

Goldberg, Plotkin and Tardos [10] constructed a $\tilde{O}(m^2n^2 \log B)$ time algorithm for generalized maximum flow problem, where $\tilde{O}(f) = O(f \log^{O(1)} n)$. This algorithm is called Fat-path algorithm. Fat-path algorithm augments a flow along to an augmenting path, and then removes the all flow generating cycles from the residual graph. If we augment the flow along to an arbitrary augmenting path, exponential time of augmentation can be needed to obtain the optimum flow. Fat-path algorithm, therefore, selects a fat path. A fat path is a path along which we can send a sufficiently large amount of flow from the source to the sink.

Radzik [12] proposed the improved version of Fat-path algorithm. This algorithm is an ε-approximate algorithm that runs in $\tilde{O}(m^2n + \min\{m^2n, m(m + n \log\log B)\} \log(1/\varepsilon))$. The main idea is that this algorithm does not remove all flow generating cycles but removes the generating cycles that have sufficiently large gains. Radzik also proposed the error-scaling algorithm where recursively solve generalized maximum flow problems with different ε.

Tardos and Wayne [13] improved the Radzik's algorithm by gain-scaling. The main idea of gain-scaling is that construct the rounded network whose gains are rounded to powers of the base $(1+\varepsilon)^{1/n}$, and calculate flow in the rounded network. This algorithm runs in $\tilde{O}(m(m+n \log\log B) \log(1/\varepsilon)+mn^2 \log B)$ time.

Fleischer and Wayne [7] proposed an ε-approximate algorithm for lossy generalized networks that runs in $O(m(m+n \log m)\varepsilon^{-2} \log n)$ time. This algorithm is based on the linear programming as well as gain-scaling and error-scaling.

Goldfarb, Jin and Orlin [6] proposed the exact algorithm that runs in $O(m^2(m+n \log n) \log B)$ time. To the best to our knowledge, this is the fastest exact algorithm for generalized maximum flow problem.

Table 1 describes the time complexity of each algorithm.

3 Preliminary

An instance of the generalized maximum flow problem is a *generalized network* $G = (V, E, s, t, \mu, \gamma)$, where V is a set of n vertices, E is a set of m edges, $s, t \in V$ are distinguished vertices called the *source* and *sink* respectively, $\mu : E \to \mathbb{R}_{\geq 0}$ is a *capacity function*, and $\gamma : E \to \mathbb{R}_{\geq 0}$ is a *gain function*. A *lossy network* is a generalized network where no edges has gain factor exceeding one. The *gain factor of a path P* is defined by $\gamma(P) := \prod_{e \in P} \gamma(e)$.

An example of generalized network is shown in Figure 3. In the figure, x/y means that x is the capacity and y is the gain factor of the corresponding edge.

A function $f : E \to \mathbb{R}_{\geq 0}$ is a *flow* of G when f satisfies the following conditions:

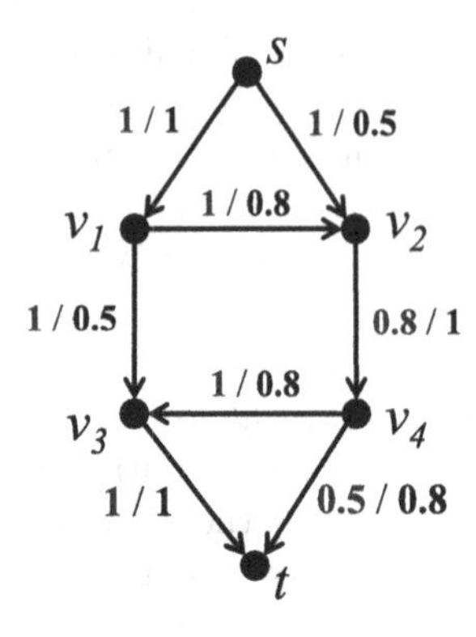

(a) A generalized network

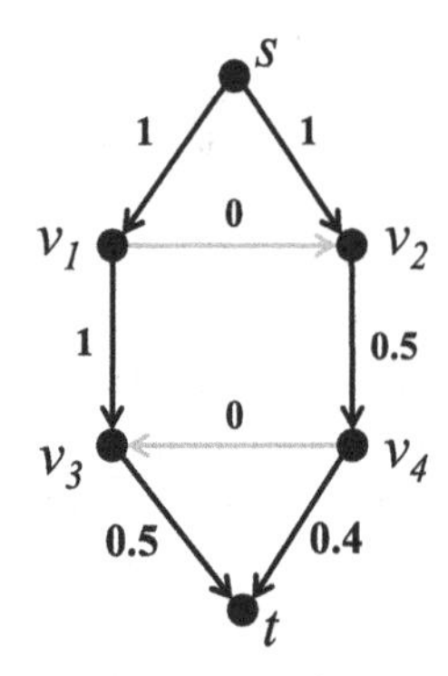

(b) The optimum flow for the generalized network depicted in Figure 3

Fig. 2. A generalized network and its optimum flow

- For each edge $e \in E$,

$$f(e) \leq \mu(e).$$

- For each vertices $v \in E \backslash \{s, t\}$,

$$\mathrm{ex}_f(v) = 0.$$

Where $ex_f(v)$ is the *residual excess* of f at a vertex v that is defined by

$$\mathrm{ex}_f(v) := \sum_{(u,v) \in E} \gamma(u,v) f(u,v) - \sum_{(v,w) \in E} f(v,w).$$

The *value* of the flow f is defined by

$$|f| := \mathrm{ex}_f(t).$$

An example of the flow is shown in Figure 3. In the figure, edges with zero flow values are illustrated by light color. The flow in the figure has a value $0.5 \cdot 1 + 0.4 \cdot 0.8 = 0.82$.

Given a real number x, flow f, and a path $P = \langle e_1, e_2, \ldots, e_k \rangle$, for each edge e_i $(i = 1, \ldots, k)$, let

$$x_i := \frac{x}{\prod_{j=i}^{k} \gamma(e_j)}.$$

Then, the augmentation of f along P by x is defined as follows: for each edge e_i $(i = 1, \ldots, k)$, augment $f(e_i)$ by x_i. Algorithm 1 is an implementation of this operation.

Algorithm 1. Augment

Input: Generalized network $G = (V, E, t, u, \gamma, \text{ex})$CFlow $f : E \to \mathbb{R}_{\geq 0}$CPath function $P : V \to \overleftrightarrow{E}$
Output: Flow of G, $f : E \to \mathbb{R}_{\geq 0}$

1: $x := 1$C$v := t$C$y := \infty$
2: **loop**
3: $\quad e := P(v)$
4: $\quad$ **if** $e =$ null **then** break
5: $\quad x \leftarrow x/\gamma(e)$
6: $\quad y \leftarrow \min\{y, u_f(e)/x\}$
7: $\quad v \leftarrow \text{tail}(e)$
8: **end loop**
9: $v \leftarrow t$
10: **loop**
11: $\quad e := P(v)$
12: $\quad$ **if** $e =$ null **then** break
13: $\quad y \leftarrow y/\gamma(e)$
14: $\quad$ **if** $e \in E$ **then**
15: $\qquad f(e) \leftarrow f(e) + y$
16: $\quad$ **else**
17: $\qquad$ Let $e' \in E$ the edge where $\overleftarrow{e'} = e$
18: $\qquad f(e') \leftarrow f(e') - y$
19: $\quad$ **end if**
20: $\quad v \leftarrow \text{tail}(e)$
21: **end loop**

4 Greedy Algorithms

4.1 Simple Greedy Algorithm

Our greedy algorithm is based on the repeated augmentation along the highest-fatness paths. The highest-fatness residual path is a path along which largest amount of flow can reach to the sink. We call this algorithm *highest-fatness greedy method* (HFG).

The highest-fatness residual path is obtained by Dijkstra-like algorithm. Dijkstra's algorithm maintains the distance $d(v)$ from s and updates the distance by

$$d(v) \leftarrow \min\{d(v), d(u) + l(u, v)\},$$

where $l(u, v)$ is length of edge (u, v). Similarly, our algorithm updates the fatness by

$$F(v) \leftarrow \max\bigl\{F(v), \gamma(u, v) \cdot \min\{F(u), \mu(u, v) - f(u, v)\}\bigr\}.$$

After the highest-fatness residual path is obtained, we augment the flow along the path as much as possible. We repeat this augmentation process until there are no s-t residual paths.

After one augmentation of a flow, at least one edge is saturated. In this algorithm, the flow value of every edge never decreases. Therefore, the number of flow

Algorithm 2. Highest-Fatness Greedy Method

Input: Lossy generalized network $G = (V, E, t, u, \gamma, \mathrm{ex})$
Output: Flow of G, $f : E \to \mathbb{R}_{\geq 0}$
1: For each $e \in E$, let $f := 0$
2: **loop**
3: For each $v \in V$, let $P(v) := \mathrm{null}$
4: For each $v \in V$, let $X(v) := \mathrm{ex}(v)$
5: $Q := \{(\mathrm{ex}(v), v, \mathrm{null}) \mid v \in V, \mathrm{ex}(v) > 0\}$
6: **while** $Q \neq \{\}$ **do**
7: Extract tuple (x, v, e) from Q that has maximum x in Q
8: **if** $P(v) \neq \mathrm{null}$ **then** continue
9: $P(v) \leftarrow e$
10: **if** $v = t$ **then** break
11: **for all** $e \in \delta^+(v)$ **do**
12: $w := \mathrm{head}(e)$
13: $x' := \gamma(e) \cdot \min\{X(v), u_f(e)\}$
14: **if** $x' > X(w)$ **then**
15: $X(w) \leftarrow x'$
16: $Q \leftarrow Q \cup \{(x', w, e)\}$
17: **end if**
18: **end for**
19: **end while**
20: **if** $P(t) = \mathrm{null}$ **then** break
21: $f \leftarrow \mathrm{Augment}(G, f, P)$
22: **end loop**

augmentation is at most m times. The time of solving a shortest path problem is dominant on the time of a flow augmentation. Therefore, the time complexity of HFG is $O(mT_{\mathrm{SP}})$, where T_{SP} is a time required for solving a shortest path problem. If we use the Dijkstra's algorithm implemented with the Fibonacci heap, T_{SP} is $O(m + n \log n)$.

4.2 Improvement by Initial Flow Calculation

Our greedy algorithms sometimes output a flow with lower value than the optimum flow. We propose an improvement by calculating a initial flow.

We first construct the instance G' from instance G of the generalized maximum flow problem. G' is defined by (V, E, s, t, μ'), where $\mu'(u, v) = \mu(u, v)/\gamma(P(u))$, and $P(u)$ is highest-gain path from s to u.

After the maximum flow f' on G' is obtained, we translate f' into initial generalized flow f on G. Without loss of generality, we assume that f' has no cycles. The translation from f' to f is defined as follows:

1. $f(e) := 0$ if $f'(e) = 0$.

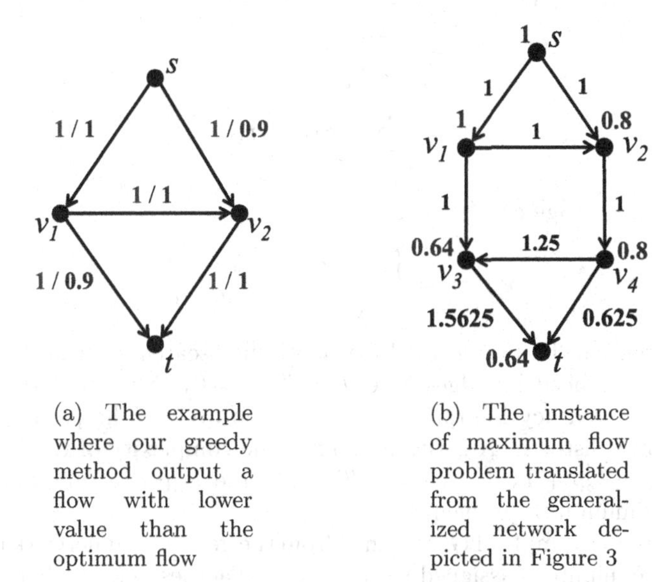

(a) The example where our greedy method output a flow with lower value than the optimum flow

(b) The instance of maximum flow problem translated from the generalized network depicted in Figure 3

Fig. 3. A failure case of the greedy method and a translated maximum flow network

---

Algorithm 3. Translate Flow

Input: Vertices V, Edges E, Gains $\gamma : E \to \mathbb{R}_{\geq 0}$CMaximum flow $f' : E \to \mathbb{R}_{\geq 0}$, Generalized flow $f : E \to \mathbb{R}_{\geq 0}$, Visited vertices $S \subset V$, Vertex $v \in V$
Output: Flow $f : E \to \mathbb{R}_{\geq 0}$
1: **if** $v \in S$ **then** return
2: $S \leftarrow S \cup \{v\}$
3: $x' := 0$C$x := 0$
4: **for all** $e \in \delta^-(v)$ **do**
5:   **if** $f'(e) > 0$ **then**
6:     $w := \text{tail}(e)$
7:     $f \leftarrow \text{TranslateFlow}(V, E, \gamma, f', f, S, w)$
8:     $x' \leftarrow x' + f'(e)$
9:     $x \leftarrow x + \gamma(e)f(e)$
10:   **end if**
11: **end for**
12: **for all** $e \in \delta^+(v)$ **do**
13:   **if** $f'(e) > 0$ **then**
14:     **if** $x' = 0$ **then**
15:       $f(e) \leftarrow f'(e)$
16:     **else**
17:       $f(e) \leftarrow xf'(e)/x'$
18:     **end if**
19:   **end if**
20: **end for**

---

2. For each vertex v, let

$$x'_v := \sum_{(u,v)\in E} f'(u,v),$$

$$x_v := \sum_{(u,v)\in E} \gamma(u,v)f(u,v).$$

Then, for each edge $e = (v,w) \in E$,

$$f(e) := \begin{cases} x_v f'(e)/x'_v & x'_v \neq 0 \\ f'(e) & x'_v = 0 \end{cases}$$

This translation can be performed by a depth-first search starting at t on the subgraph of G induced by edges $\{e \in E \mid f'(e) > 0\}$. Note that the induced subgraph is directed acycle graph since f' has no cycles. The pseudo-code for the translation is listed in Algorithm 3. The time complexity of the initial flow calculation is $O(T_{\mathrm{SP}} + T_{\mathrm{MF}} + m) = O(T_{\mathrm{MF}})$, where T_{MF} is a time required for solving a maximum flow problem.

Figure 4.2 is an example of G' obtained from the generalized network in Figure 3. In the figure, numbers assigned to edges are capacities and numbers assigned to vertices are $\gamma(P(u))$.

After the translation, we augment the initial flow f using the same method as that used in HFG. Since this heuristic does not always give the better approximation than simple HFG, we first calculate both simple HFG and HFG with the initial flow, then select the flow with larger value. We call this algorithm HFG-M.

5 Experiment

In respect to computation times and values of flows, we compared our algorithms with Fleischer and Wayne's ε-approximation algorithm with $\varepsilon = 0.2$

We used a computer with Core i5-2500K 3.3 GHz, 16 GB memory. Our program was implemented by C#.

The input network is the Japanese Wikipedia at January 21, 2012. We view the Wikipedia as the information network whose vertices are corresponding to the articles and whose edges are corresponding to links between articles. The number of vertices was 2,087,085 and the number of edges was 63,230,008. Because the input network is too large to calculate the solution with the baseline algorithm, we sampled the network with random walk sampling introduced by Leskovec and Faloutsos [15]. The obtained network has 100,000 vertices and 268,950 edges. We used this sampled network for our experiment.

In our experiment, we utilized two kinds of capacities and gain factors as follows.

- Fixed capacities and gain factors ($u(e) \equiv 1$, $\gamma(e) \equiv 0.5$).
- Randomized capacities and gain factors (chosen uniformly within $0 \leq u(e) \leq 2C0 \leq \gamma(e) \leq 1$).

Table 2. Ratio of flow values and times

			Fixed	Random
HFG / FW	Flow value	Average	1.00	0.97
		Worst	1.00	0.19
	Speed up	Average	527	562
HFG-M / FW	Flow value	Average	1.00	1.00
		Worst	1.00	0.31
	Speed up	Average	157	146

Fifty pairs of vertices are randomly selected for the sources and sinks.

With this generalized network, we computed the generalized maximum flow using HFG and Fleischer-Wayne's method (FW). Then, for each flow, we calculated the ratio between two method of computation time and the ratio between two method of flow value. Finally we calculated the average computation time ratio and average flow value ratio.

Experimental result are shown in table 2. The HFG is more than 500 times faster than FW. The values of flows obtained by the two algorithms are comparable in average. The HFG-M is about 150 times faster than FW and its worst flow value is 50% better than HFG's one.

6 Conclusion

We proposed an approximate algorithm for the generalized maximum flow problem. This algorithm is based on a greedy method where repeatedly send the flow along the highest fatness augmenting path in the graph. The experimental result showed that this algorithm is order of magnitudes faster than the Fleischer-Wayne's method with $\varepsilon = 0.2$, and has a comparable accuracy to Fleischer-Wayne's method. We believe that this algorithm sufficient for the relation extraction in large real information networks like Wikipedia.

We also proposed an initial flow calculation method to improve the accuracy of our greedy method. This method is based on the translation from generalized maximum flow problem to maximum flow problem. After obtaining the maximum flow from the source to the sink, we construct the initial flow for generalized maximum flow problem from the maximum flow. The experimental result showed that this method improves the accuracy in the worst case.

This paper discusses the practical improvements of the computation time of generalized maximum flow problem. Real complex networks often have special structures such as scale-free network and small-world network. We expect that some theoretical bounding of time complexity can be hold by assuming the such network structures.

References

1. Zhang, X., Asano, Y., Yoshikawa, M.: A generalized flow based method for analysis of implicit relationships on wikipedia. IEEE Transactions on Knowledge and Data Engineering (November 10, 2011)
2. Yazdani, M., Popescu-Belis, A.: A random walk framework to compute textual semantic similarity: A unified model for three benchmark tasks. In: ICSC, pp. 424–429 (2010)
3. Sarkar, P., Moore, A.W.: A tractable approach to finding closest truncated-commute-time neighbors in large graphs. In: UAI, pp. 335–343 (2007)
4. White, H.D., Griffith, B.C.: Author cocitation: A literature measure of intellectual structure. Journal of the American Society for Information Science 32(3), 163–171 (1981)
5. Ito, M., Nakayama, K., Hara, T., Nishio, S.: Association thesaurus construction methods based on link co-occurrence analysis for wikipedia. In: CIKM, pp. 817–826 (2008)
6. Goldfarb, D., Jin, Z., Orlin, J.B.: Polynomial-time highest gain augmenting path algorithms for the generalized circulation problem. Mathematics of Operations Research 22, 793–802 (1997)
7. Fleischer, L.K., Wayne, K.D.: Fast and simple approximation schemes for generalized flow. Mathematical Programming 91(2), 215–238 (2002)
8. Kapoor, S., Vaidya, P.M.: Speeding up karmarkar's algorithm for multicommodity flows. Mathematical Programming 73, 111–127 (1996)
9. Vaidya, P.M.: Speeding-up linear programming using fast matrix multiplication (extended abstract). In: IEEE Symposium on Foundations of Computer Science, pp. 332–337 (1989)
10. Goldberg, A.V., Plotkin, S.A., Tardos, É.: Combinatorial algorithms for the generalized circulation problem. In: IEEE Symposium on Foundations of Computer Science, pp. 432–443 (1988)
11. Cohen, E., Megiddo, N.: New algorithms for generalized network flows. In: Dolev, D., Rodeh, M., Galil, Z. (eds.) ISTCS 1992. LNCS, vol. 601, pp. 103–114. Springer, Heidelberg (1992)
12. Radzik, T.: Faster algorithms for the generalized network flow problem. In: IEEE Symposium on Foundations of Computer Science, pp. 438–448 (1993)
13. Tardos, É., Wayne, K.D.: Simple generalized maximum flow algorithms. In: Integer Programming & Combinatorial Optimization, pp. 310–324 (1998)
14. Onaga, K.: Dynamic programming of optimum flows in lossy communication nets. IEEE Trans. Circuit Theory, 308–327 (1966)
15. Leskovec, J., Faloutsos, C.: Sampling from large graphs. In: ACM SIGKDD Conference on Knowledge Discovery and Data Mining, pp. 631–636 (2006)

A Necessary and Sufficient Condition for a Bipartite Distance-Hereditary Graph to Be Hamiltonian

Masahide Takasuga and Tomio Hirata

School of Information Science, Nagoya University,
Furocho, Chikusaku, Nagoya, Japan
hirata@is.nagoya-u.ac.jp

Abstract. In this paper, we present a necessary and sufficient condition for a bipartite distance-hereditary graph to be Hamiltonian. The result is in some sense similar to the well known Hall's theorem, which concerns the existence of a perfect matching. Based on the condition we also give a polynomial-time algorithm for the Hamilton cycle problem on bipartite distance-hereditary graphs.

1 Introduction

A graph G is *distance-hereditary* if for each induced connected subgraph G' of G and for any two vertices u, v of G', the distance between u and v does not change, that is, $d_{G'}(u, v) = d_G(u, v)$, where $d_G(u, v)$ is the length of a shortest path in G between u and v. Distance-hereditary graphs form a subclass of perfect graphs. Two well-known classes of graphs, trees and cographs, are both subclasses of distance-hereditary graphs.

For a undirected graph $G = (V, E)$, a Hamilton cycle of G is a cycle which passes each vertex of G exactly once. A graph is said to be Hamiltonian if it contains a Hamilton cycle. The Hamilton cycle problem is a problem to decide whether a given graph is Hamiltonian or not.

In this paper we present a necessary and sufficient condition for a bipartite distance-hereditary graph (BDHG) to be Hamiltonian, and give a polynomial-time algorithm for the Hamilton cycle problem on this class of graphs. Note that the Hamilton cycle problem on chordal bipartite graph, which is a superclass of the bipartite distance-hereditary graphs, is known to be NP-complete[13].

Our polynomial time algorithm for BDHG is not the first one. Müller and Nicolai[12] proposed the first polynomial time algorithm in 1993. Since then there have been incremental improvements, and eventually resulting in linear time algorithms [14,7,6,8]. [1] However, these algorithms are quite complicated and difficult to implement. Our algorithm is based on the necessary and sufficient condition we propose in this paper, and easy to implement.

[1] Actually, these improvements and linear time algorithms are for distance-hereditary graphs, not restricted to BDHG.

J. Akiyama, M. Kano, and T. Sakai (Eds.): TJJCCGG 2012, LNCS 8296, pp. 143–149, 2013.

The rest of this paper is organized as follows. In Section 2, we give some basic definitions and review some properties of distance-hereditary graphs. Section 3 introduces the reduced graph. In Section 4, we present a necessary and sufficient condition for a bipartite distance-hereditary graph to be Hamiltonian. Finally, in Section 5, we present a polynomial-time algorithm for the Hamilton cycle problem on bipartite distance-hereditary graphs.

2 Preliminaries

This paper considers a finite, simple and undirected graph $G = (V, E)$, where V and E are the vertex and edge sets of G, respectively. For a vertex x of G, we denote by $N_G(x)$ the set of adjacent vertices of x. We write $N(x)$ when the underlying graph is clear. A graph $G = (V, E)$ is bipartite if V can be partitioned into two subsets V^+ and V^- such that any edge of G has one terminal in V^+ and the other terminal in V^-. A bipartite graph may be denoted by $G = (V^+, V^-, E)$. A bipartite graph $G = (V^+, V^-, E)$ is complete if every vertex of V^+ is adjacent to every vertex in V^-. A graph is chordal if any cycle of length ≥ 4 has a chord. A graph is (k, l)-chordal if any cycle of length $\geq k$ has at least l chords.

Bandelt and Mulder[1] showed that every connected distance-hereditary graph G can be generated by iterating a *one-vertex extension* from K_1. The one-vertex extension is a set of three operations to add a new vertex to a graph: adding pendant vertex operation (PV) and two twin operations (FT and TT). Specifically, for $G = (V, E)$ and $x \in V$, we add a new vertex $x' \notin V$ as follows: $V \leftarrow V \cup \{x'\}$ and
$E \leftarrow E \cup \{(x, x')\}$ (PV),
$E \leftarrow E \cup \{(x', y) | y \in N(x)\}$ (FT) or
$E \leftarrow E \cup \{(x', y) | y \in N(x) \cup \{x\}\}$ (TT).

In this paper, we consider only connected graphs. The following (1),(2),(3) are equivalent[1].

(1) G is a BDHG.
(2) G can be generated from K_1 by iterating PV and FT operations.
(3) G is $(6, 2)$-chordal bipartite.

3 Reduced Graph

For a BDHG $G = (V^+, V^-, E)$, we partition each V^+ and V^- into groups and make a new graph G^r so that each vertex of G^r represents a group of G. Let R be the equivalence relation on $V(= V^+ \cup V^-)$ such that ${}_{v_i}R_{v_j}$ iff $N(v_i) = N(v_j)$. We partition V into the equivalence classes $\gamma_i (1 \leq i \leq t)$ of R. From the equivalence classes, we construct a graph $G^r = (V^r, E^r)$, the *reduced graph*, as follows. Each vertex v_i of G^r represents γ_i, and if there is an edge between a vertex of γ_i and a vertex of γ_j, we add an edge (v_i, v_j) into E^r.

In terms of the modular decomposition[16], our equivalence class corresponds to a module of $G = (V^+, V^-, E)$. For a undirected graph $G = (V, E)$, a set $M \subseteq V$ is a *module* of G if $N(x) \setminus M = N(y) \setminus M$ holds for all vertices x and y in M. The modular decomposition of a graph is a recursive (hierarchical) decomposition of the graph into modules. For a graph G, this decomposition is unique and represented by a rooted tree whose nodes correspond to modules of G; the root to V , the leaves to single vertices, and each parent module contains all its child modules. The modular decomposition of a graph can be computed in linear time[3,5,11,15]. In case of bipartite graphs, the modular decomposition is just a partition of the vertices of G and the rooted tree is of hight 2. Our reduced graph G^r is called *characteristic graph* in the modular decomposition.

The following observation is essential in our discussion.

Proposition 1. *Let G be a* BDHG *with at least one edge. Then there is at least one vertex of degree* 1 *in its reduced graph G^r.*

Proof. We show the proposition by an induction on the number n of vertices of G. When $n = 2, G$ has two vertices of degree 1. Let G be a BDHG of $n+1$ vertices and assume that G is constructed by adding a new vertex x' to a vertex x of a graph G' by a PV operation. The equivalence class including x' is a singleton set, and hence the reduced graph G^r has at least one vertex of degree 1. If G is constructed by adding a vertex x' to a vertex x of G' by an FT operation. Then the reduced graph does not change by this operation, since the added vertex x' belongs to the equivalence class of x in G. From the induction hypothesis, the reduced graph of G' has a vertex of degree 1, and hence the reduced graph of G has a vertex of degree 1. □

4 Expanding Condition

Assume that $G = (V^+, V^-, E)$ is a bipartite graph with $2n$ vertices and $|V^+| = |V^-| = n$. Let X be any vertex set of V^+ with $|X| < n$. We denote by $N(X)$ the set of adjacent vertices of vertices in X, that is, $N(X) = \bigcup_{v \in X} N(v)$. It is clear that if G has a Hamilton cycle, $|X| < |N(X)|$. We call the following condition an *expanding condition for* V^+ of G.

$$\forall X \subsetneq V^+ \quad |X| < |N(X)| \tag{1}$$

For any bipartite graph $G = (V^+, V^-, E)$ with $|V^+| = |V^-|$, the expanding condition for V^+ holds if and only if the expanding condition for V^- holds.

Proposition 2. *The following condition is equivalent to (1).*

$$\forall X \subsetneq V^- \quad |X| < |N(X)| \tag{2}$$

Proof. We show that (1) implies (2). Assume that (2) does not hold for some $X \subsetneq V^-$, that is, $|X| \geq |N(X)|$ for this X. Since $|V^+| = |V^-|$, we have $|V^- - X| \leq |V^+ - N(X)|$. Since there is no edge between X and $(V^+ - N(X))$, we have

$N(V^+ - N(X)) \subset (V^- - X)$. Therefore, $|V^+ - N(X)| \geq |N(V^+ - N(X))|$. This means that (1) does not hold. The reverse direction can be proved in the same manner. □

It is obvious that the expanding condition is necessary for G to have a Hamilton cycle. However, it is not a sufficient condition if G is a general bipartite graph. See Fig.1. In the rest of this section, we will show that if G is a BDHG then the expanding condition is a sufficient condition for G to have a Hamilton cycle.

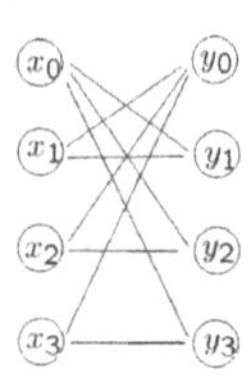

Fig. 1. A non-Hamiltonian bipartite graph satisfying the expanding condition

Let G be a BDHG and v_i a vertex of degree 1 in the reduced graph G^r. Let x be a vertex of G belonging to the equivalence class γ_i, and $y \in N(x)$. We denote by $G \backslash \{x, y\}$ the graph obtained by deleting x and y from G. Then, the following lemmas hold.

Lemma 1. *If the expanding condition holds for G, then $G \backslash \{x, y\}$ is also a connected BDHG.*

Proof. Since G is a BDHG, G is $(6, 2)$-chordal. Deleting x and y from G does not delete a chord of a cycle in G and it does not produce a new cycle in G. Therefore the resulted graph is also $(6, 2)$-chordal and bipartite. Furthermore, since the expanding condition holds for G, $|N_G(x)| \geq 2$, and hence $G \backslash \{x, y\}$ is connected.

Lemma 2. *Assume that the expanding condition holds for G. If $G \backslash \{x, y\}$ is Hamiltonian, G is also Hamiltonian*

Proof. Let H be a Hamilton cycle of $G \backslash \{x, y\}$. Let $z(\neq y)$ be a vertex of $N_G(x)$ and $u(\neq x)$ is an adjacent vertex of z such that H passes through edge (u, z). The expanding condition guarantees the existence of such z and u. Since v_i is of degree 1 in G^r, z and y belong to the same equivalence class, that is $N(z) = N(y)$. Hence there exists an edge between u and y. Thus we can modify H so that it passes through vertices u, y, x, z in this order. We have a Hamilton cycle of G. □

Lemma 3. *The expanding condition holds for $G \backslash \{x, y\}$ whenever it holds for G.*

Proof. We will show that for any $X \subsetneq (V^+ - \{x\})$

$$|X| < |N_{G \setminus \{x,y\}}(X)|. \tag{3}$$

Since the expanding condition holds for G, we have $|X| < N_G(X)$ with respect to this X. Hence (3) obviously holds if $y \notin N_G(X)$.

We consider the case $y \in N_G(X)$. Let $X' = X \cup \{x\}$. Since $X' \subsetneq V^+$, the expanding condition for G implies $|X'| < |N_G(X')|$. Let z be any vertex of X adjacent to y. The expanding condition for G guarantees the existence of such z. Since z is adjacent to each vertex of $\gamma[y]$ (the equivalence class of y) and x is adjacent exactly to the vertices of $\gamma[y]$, $N_G(X') = N_G(X)$. Therefore,

$$|X| < |X'| < |N_G(X')| = |N_G(X)|$$

This implies (3). □

From the above lemmas, we have the following theorem.

Theorem 1. *Let $G_{2n} = (V^+, V^-, E)$ be a* BDHG *of $2n$ vertices and $|V^+| = |V^-| = n(\geq 2)$. If the expanding condition holds for G_{2n}, there is a Hamilton cycle in G_{2n}.*

Proof. We show the theorem by an induction on the value of n.

If the graph has 4 vertices, from the expanding condition G_{2n} is a biclique and a Hamilton cycle exists. We assume that the theorem holds for G_{2n} and show it also holds for $G_{2(n+1)}$. Since $G_{2(n+1)}$ is a BDHG, from Proposition 1 there is a vertex v_i of degree 1 in $G^r_{2(n+1)}$. From Lemma 3, the expanding condition holds for $G_{2(n+1)} \setminus \{x, y\}$ that is obtaind from $G_{2(n+1)}$ by deleting $x \in \gamma_i$ and $y \in N(x)$. From Lemma 1 and the induction hypothesis, there is a Hamilton cycle in $G_{2(n+1)} \setminus \{x, y\}$. Therfore, from Lemma 2, $G_{2(n+1)}$ has a Hamilton cycle. □

Chvátal introduced the concept of toughness, a measure of connectivity that is closely connected to the existence of Hamilton cycles[4,2] . For a graph $G = (V, E)$ let $c(G)$ denote the number of its connected components. The *toughness* of G is the minimum value of $|S|/c(G - S)$ over all sets S such that $\emptyset \subset S \subset V$. It is easy to see that if G is Hamiltonian then $t(G) \geq 1$ and if G is bipartite then $t(G) \leq 1$. Thus if G is bipartite and Hamiltonian then $t(G) = 1$. We have the following corollary.

Corollary 1. *For a BDHG G, $t(G) = 1$ implies that G is Hamiltonian.*

Proof. For a BDHG G, if $t(G) = 1$ then the expanding condition holds. Consider to the contrary that the expanding condition does not hold. Then there is a subset $X \subsetneq V^+$ with $|X| \geq |N(X)|$. Deleting the vertices of $N(X)$ from G yields more than $|X|$ components, contradicting that $t(G) = 1$. Then our theorem implies that G is Hamiltonian. □

Note that there are non-Hamiltonian bipartite graphs with $t(G) = 1$. See Figure 1 in [9]. Indeed, that graph is not a BDHG.

5 Algorithm

Based on the reduced graph and the lemmas above, we can construct a polynomial time algorithm for deciding whether a BDHG $G = (V^+, V^-, E)$ is Hamiltonian.

procedure *Hamilton*(G)
1. Construct the reduced graph G^r of G.
2. If G^r consists of a single edge,
   output "G is Hamiltonian" and stop.
3. Let v_i be a vertex of G^r with $\deg(v_i)$=1 and v_j its adjacent vertex.
   Let γ_i and γ_j be the corresponding equivalence classes.
4. If $|\gamma_i| \geq |\gamma_j|$, output "$G$ is not Hamiltonian" and stop.
5. Delete the vertices of γ_i and the $|\gamma_i|$ vertices of γ_j from G.
   Let G be the resulted graph and goto Step 1.

This algorithm can easily be extended to become *certifying*(see [10]): When it rejects in Step 4 the sets γ_i and γ_j certify that the current graph does not satisfy the expanding condition. Also when the algorithm terminates in Step 2, we can trace back the graphs appeared in the algorithm to produce the Hamilton cycle.

Although polynomial-time algorithms have been known for BDHG as noted in Introduction, our algorithm is simple and implementation would be easy. A naive implementation yields an $O(mn)$ time algorithm, where n is the number of vertices and m is the number of edges in the input graph. We can use linear-time modular decomposition algorithms[3,5,11,15] for Step 1 and Step 1 may be repeated $O(n)$ times, yielding an $O(mn)$ time algorithm. Introducing an elaborate data structure, we can implement this algorithm so that it runs in linear time. We do not go into the discussion of implementation, since it is beyond the scope of this paper.

Acknowledgment. The authors would like to thank the anonymous referees for their helpful comments, which improve the readability of the original draft.

References

1. Bandelt, H.-J., Mulder, H.M.: Distance-Hereditary Graphs. Journal of Combinatorial Theory B 41, 182–208 (1986)
2. Bauer, D., Broersma, H., Schmeichel, E.: Toughness in Graphs A Survey. Graphs and Combinatorics 22, 1–35 (2006)
3. Cournier, A., Habib, M.: A new linear algorithm for modular decomposition. In: Tison, S. (ed.) CAAP 1994. LNCS, vol. 787, pp. 68–84. Springer, Heidelberg (1994)
4. Chvátal, V.: Tough graphs and Hamiltonian circuits. Discrete Mathematics 5, 215–228 (1973)
5. Habib, M., de Montgolfier, F., Paul, C.: A simple linear-time modular decomposition algorithm for graphs, using order extension. In: Hagerup, T., Katajainen, J. (eds.) SWAT 2004. LNCS, vol. 3111, pp. 187–198. Springer, Heidelberg (2004)

6. Hsieh, S.-Y., Ho, C.-W., Hsu, T.-S., Ko, M.-T.: The problem on distance-hereditary graphs. Discrete Applied Mathematics 154, 508–524 (2006)
7. Hung, R.-W., Wu, S.-C., Chang, M.-S.: Hamiltonian cycle problem on distance-hereditary graphs. Journal of Information Science and Engineering 19, 827–838 (2003)
8. Hung, R.-W., Chang, M.-S.: Linear-time algorithms for the Hamiltonian problems on distance-hereditary graphs. Theoretical Computer Science 341, 411–440 (2005)
9. Kratsch, D., Lehel, J., Müller, H.: Toughness, hamiltonicity and split graphs. Discrete Mathematics 150, 231–245 (1996)
10. McConnell, R.M., Mehlhorn, K., Näher, S., Schweitzer, P.: Certifying Algorithms. Computer Science Review 5, 119–161 (2011)
11. McConnell, R.M., Spinrad, J.P.: Linear-time modular decomposition and efficient transitive orientation of comparability graphs. In: Proceeding SODA, pp. 536–545 (1994)
12. Müller, H., Nicolai, F.: Polynomial time algorithms for Hamiltonian problems on bipartite distance-hereditary graphs. Information Processing Letters 46, 225–230 (1993)
13. Müller, H.: Hamiltonian circuits in chordal bipartite graphs. Discrete Mathematics 156, 291–298 (1996)
14. Nicolai, F.: Hamiltonian problems on distance-hereditary graphs, Technique Report SM-DU-264, Gerhard-Mercator University, Germany (1994)
15. Tedder, M., Corneil, D.G., Habib, M., Paul, C.: Simpler linear-time modular decomposition via recursive factorizing permutations. In: Aceto, L., Damgård, I., Goldberg, L.A., Halldórsson, M.M., Ingólfsdóttir, A., Walukiewicz, I. (eds.) ICALP 2008, Part I. LNCS, vol. 5125, pp. 634–645. Springer, Heidelberg (2008)
16. Möhring, R.H., Radermacher, F.J.: Substitution decomposition for discrete structures and connections with combinatorial optimization. Annals of Discrete Mathematics 19, 257–356 (1984)

On Simplifying Deformation of Smooth Manifolds Defined by Large Weighted Point Sets

Ke Yan and Ho-Lun Cheng

National University of Singapore
{yanke2,hcheng}@comp.nus.edu

Abstract. We present a simple and efficient algorithm for deformation between significantly different objects, which does not require any forms of similarity or correspondence. In our previous work, the algorithm, which is called general skin deformation, requires a complexity of $O(m^2n^2)$. In this paper, we improve the complexity from $O(m^2n^2)$ to $O(m^2 + n^2)$ by proposing a simplified deformation process. This improvement greatly reduces the program running time and unnecessary topology changes. Moreover, it makes some impossible deformation with large input sets to become possible.

1 Introduction

Almost all available deformation techniques require similarity to establish correspondence mapping of the source and target shapes, e.g. deformations between different postures of human or animals [1,3,12,15,16]. For deformation between significantly different objects, e.g. a mannequin head and a fist (Figure 1), it is always a challenge to create a smooth transition from one to another. These examples are not unreasonable since many breath-taking movies and cartoons require deformation animations between rather different objects (e.g. the liquid robot deforms in the movie Terminator 2). Automatic correspondence mapping methods suffer in these cases, and usually require heavy user labor. Two problems arise for deformation between non-similar objects: 1. there are ambiguities in vertex mapping between the source and target shapes 2. there are difficulties to handle topology changes automatically.

In 2010, we propose a general deformation process for any shapes approximated by a smooth manifolds called *skin surfaces* [9]. The general skin deformation (GSD) requires no correspondence mapping information and handles topology changes automatically. We convert the source and target shapes into weighted point sets B_0 and B_1, by existing algorithms such as the power crust [2] or the sphere-tree toolkit [4]. The general skin surface deformation algorithm generates skin surfaces skin (B_0) and skin (B_1) and deforms one skin surface into another with a parameter $t \in [0..1]$ as time.

However, the complexity of GSD increases with the sizes of input weighted point sets. Given two input weighted point sets with m and n points, an intermediate complex is required to be constructed by mn weighted points. Due to

J. Akiyama, M. Kano, and T. Sakai (Eds.): TJJCCGG 2012, LNCS 8296, pp. 150–161, 2013.

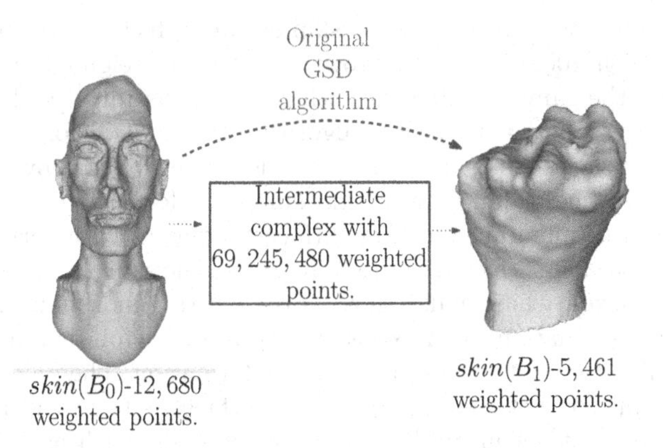

Fig. 1. It is hard to deform between a mannequin head and a fist because of the large number of intermediate weighted points

the complexity of Delaunay triangulation, the overall complexity is $O(m^2n^2)$. This quadratic complexity slow down the entire deformation process because all mesh points have to float through $O(m^2n^2)$ numbers of blocks which we called intermediate *mixed cells* [7]. The deformation of skin surface over such a dense intermediate mixed cell structure results heavy computation of *scheduling* [9], which is directly associated with the overall computation time of GSD. In Figure 1, we deform a mannequin head model skin (B_0) with 12,680 weighted points to a fist model skin (B_1) with 5,461 weighted points. The intermediate complex consists of a number of 69 million of weighted points. The complexity of this intermediate complex is the square of 69 million which is impossible to be handled with a 32-bit machine.

In this paper, we present a simple and efficient deformation solution, namely, SGSD, for drastically different source and target objects with no correspondence mapping information provided. We inherit most features from our previous work (GSD) and improve the overall complexity from $O(m^2n^2)$ to $O(m^2 + n^2)$. This improvement greatly reduces the program running time. Moreover, it abandons unnecessary topology changes and makes some impossible deformations with large input sets become possible. New degeneracy problems arise in the simplified deformation process and we solve them by introducing new types of intermediate complexes.

1.1 Related Work

The skin surface is firstly introduced as a maximum curvature continuous surface model for molecules by Edelsbrunner in 1999 [10]. It has several distinct properties such as smoothness, deformability and complementary, which are desirable in biological studies such as protein docking and protein-protein interactions [13].

The first deformation framework of skin surface which is the growth model, is implemented in 2002 [7]. In the growth model, all weighted points increase or decrease by the same α value [11], which is not useful in real world applications. For years, there is no proper algorithm to handle deformation between arbitrary skin surfaces, although the initial idea was proposed by Edelsbrunner in 1999 [10]. The main difficulty of skin surface deformation is all intermediate shapes have to maintain the skin surface properties with correct topology. The intermediate skin surfaces require to maintain the same homology groups as the alpha complexes, which makes most of the existing morphing theories, such as the shape-interpolation by Alexa et al. [1] and skeleton-driven deformation works [15], failed to handle the skin surface deformation. The only way to visualize the deformation between different skin surfaces is to generate each frame separately using static skin mesh generation methods, such as CGAL [5] and quality skin mesh software developed by Cheng and Shi [8]. These methods has several disadvantages, such as lack of efficiency, no surface point correspondence and discontinuous of homology group changes [9]. In 2006, Cheng and Chen find that the super-imposed Voronoi diagram of two or more skin surfaces remains unchanged during the deformation process [6]. This makes continuous skin deformation possible for any combination of skin surfaces. In 2010, we implement GSD algorithm to deform between any given skin surfaces under general position assumption (GPA) [9].

The interest of skin surface deformation is no longer restricted in molecular studies but applied to all forms of objects that are representable by sets of weighted points. With the help from algorithms converting polygonal objects into weighted point sets [2,4,14], the GSD algorithm can perform global deformation between real world objects approximated by skin surfaces. The advantage of GSD algorithm over other deformation technologies is that the GSD algorithm requires no similarity or correspondence mapping information about the source and target shapes. The GSD software is available at http://www.comp.nus.edu.sg/~yanke2/skin/skin.htm.

2 Skin Surface Deformation

In this section, we briefly introduce the basics of the GSD algorithm. For the details of the implementation, including experimental results, readers can refer to our previous works [6,7,9].

2.1 Delaunay, Voronoi Complexes under General Position Assumption

A *weighted point* in $\mathbb{R}^d$ can be written as $b_i = (z_i, w_i) \in \mathbb{R}^d \times \mathbb{R}$ where $z_i \in \mathbb{R}^d$ is the position and $w_i \in \mathbb{R}$ is the weight. A weighted point is also viewed as a ball with center z_i and radius $\sqrt{w_i}$. Given a finite set $B = \{b_1, b_2, ..., b_n\}$, $z(B) \subset \mathbb{R}^d$ is denoted as the set of the positions of the weighted points in B. The weighted distance of a point $x \in \mathbb{R}^d$ from a weighted point b_i is defined as

$\pi_{b_i}(x) = \|xz_i\|^2 - w_i$. The *Voronoi region* ν_i for each weighted point $b_i \in B$ is defined as,

$$\nu_i = \{x \in \mathbb{R}^d \mid \pi_{b_i}(x) \leq \pi_{b_j}(x), b_j \in B\}.$$

For a set of weighted points $X \subseteq B$, the *Voronoi cell* of X is defined as $\nu_X = \bigcap_{b_i \in X} \nu_i$. The collection of all the non-empty Voronoi cells is called the *Voronoi complex* of B, denoted as, V_B. For each $\nu_X \in V_B$, its corresponding *Delaunay cell*, δ_X, is the convex hull of the set of centers of X, namely, conv $(z(X))$, where $\nu_X \neq \phi$. The collection of all the Delaunay cells is called the *Delaunay complex* of B, denoted as D_B.

Usually, a general position assumption is made so that the Delaunay complex is simplicial, namely, $\forall \nu_X \in V_B$, $card(X) = dim(\delta_X)+1$. Under this assumption, there are only four types of Delaunay cells in $\mathbb{R}^3$: vertices, edges, triangles and tetrahedra.

2.2 Intermediate Voronoi Complexes

During the deformation process, a series of intermediate skin surface meshes skin $(B(t))$ is generated from skin (B_0) to a new skin surface skin (B_1). In our previous work, we prove that for any $t \in [0, 1]$, all weighted point sets $B(t)$ share the same Voronoi complex $V(t)$, which we call the intermediate Voronoi complex [6]. This intermediate Voronoi complex can be obtained by super-imposing the two Voronoi diagrams of B_0 and B_1. Let the Voronoi complex of B_0 be V_0, and the Voronoi complex of B_1 be V_1. This intermediate Voronoi complex is the super-imposition of V_0 and V_1,

$$V(t) = \{\nu_{XY} | \nu_X \in V_0, \nu_Y \in V_1, \nu_X \cap \nu_Y \neq \emptyset\}.$$

For $t \in [0, 1]$, with the invariance of the intermediate Voronoi complex, we determine the type of intermediate Voronoi cells in the process of deformation. We assume the two weighted point sets are under GPA individually. In $\mathbb{R}^d$, the dimension of $\nu_{XY} = \nu_X \cap \nu_Y$, $\nu_X \in V_0$ and $\nu_Y \in V_1$, is

$$dim(\nu_{XY}) = dim(\nu_X) + dim(\nu_Y) - d. \tag{1}$$

In $\mathbb{R}^3$, after super-imposing Voronoi complexes, there are in total six possibilities of intermediate Voronoi cell types. Each type of intermediate Voronoi cells is classified by a tuple, namely, $(dim(\nu_X)$,$dim(\nu_Y)$,$dim(\nu_{XY}))$. We assume that $dim(\nu_X) > dim(\nu_Y)$ and all possible tuples are (3,3,3), (3,2,2), (3,1,1), (3,0,0), (2,2,1) and (2,1,0).

2.3 Intermediate Delaunay Complexes

We denote the intermediate Delaunay complex, $D(t)$, as the Delaunay complex of $B(t)$, and it is not a simplicial complex. Apart from regular Delaunay triangulation, we define the intermediate Delaunay complex as,

$$D(t) = \{conv(z(\nu^{-1}(\nu_{XY})))|\nu_{XY} \in V(t)\},$$

where

$$z(X) = \{z_i | b_i \in X\}, and$$
$$\nu^{-1}(\nu_{XY}) = \{b(t) | \nu_X \in V_0, \nu_Y \in V_1, \nu_X \cap \nu_Y \neq \emptyset\}.$$

2.4 Skin Decomposition

The skin can be decomposed by mixed cells. A mixed cell μ_X is the Minkowski sum of a Delaunay cell and its corresponding Voronoi cell, formally $\mu_X = (\delta_X + \nu_X)/2$. The center and size of a mixed cell are defined as

$$z_X = \text{aff}(\delta_X) \cap \text{aff}(\nu_X), and$$
$$w_X = w_i - \|z_X z_i\|^2.$$

where $b_i = (z_i, w_i)$ is any weighted point in X.

Within each mixed cell μ_X, skin $(B) \cap \mu_X$ is a quadratic surface. In $\mathbb{R}^3$, skin patches are pieces of spheres and hyperboloids of revolution.

2.5 Mesh Point Movement and Escaping Time Scheduling

We apply linear interpolation from every weighted point $b_i \in B_0$ to every weighted point $b_j \in B_1$, namely, $b_{ij}(t) = (1-t) \cdot b_i + t \cdot b_j$. Each mixed cell center z_X moves linearly since ν_X is fixed and δ_X moves linearly with $b(t)$. Skin patches expand or shrink according to the update of w_X. We triangulate skin surfaces skin $(B(t))$ with meshes and move the mesh surface points.

Scheduling is a technique while we trigger a special event in the deformation process. While a scheduled time reaches, the deformation pauses to execute the scheduled event and resume afterwards. The total number of schedules directly affects the efficiency of the program.

Every surface vertex p has an escaping time schedule t_p which is the time while p escapes from its mixed cell. The program moves p to a new mixed cell at time t_p and gives another escaping time schedule to p in the new mixed cell. The total number of the escaping time schedules depends on the complexity of the intermediate complex. Therefore the original GSD algorithm suffers from the problem of too many intermediate mixed cells.

3 Simplified General Skin Deformation Algorithm

We propose a simplified GSD (SGSD) algorithm to improve the efficiency and the complexity of intermediate complex in GSD. Given two skin surfaces, skin (B_0) with m weighted points, and skin (B_1) with n weighted points, we improve the overall complexity of GSD from $O(m^2n^2)$ to $O(m^2 + n^2)$ by three steps:

1. **Simplify Weighted Point Set.** We consider the input weighted point sets as unions of balls and simplify the unions of balls with the number of simplified weighted points given by user. We generate skin surfaces for both the original weighted point sets and the simplified weighted point sets as inputs for SGSD algorithm.

2. **Deform Skin Surfaces.** The whole SGSD process is divided into three parts. We name the three parts as Deformation I, II and III, for deformation from skin (B_0) to skin (B_0'), from skin (B_0') to skin (B_1') and from skin (B_1') to skin (B_1) respectively.
3. **Connect the Three Deformation Processes.** The whole deformation process is constructed by connecting Deformation I, II and III. For later deformations like Deformation II and III, we make use of the result from the previous deformation. The mesh points therefore move continuously in between of different deformations.

3.1 Simplifying the Union of Balls

Given a large set of balls, our goal is to simplify it into a reasonable small set with m' (given by the user) balls with a similar volume. By "reasonable" we mean that we do not have to set a tight bound on the difference between the volumes of the two unions of balls because the simplified skin surface is one of the intermediate shapes in the deformation sequence.

There are two repeating steps in simplifying the balls, namely, deletion and enlarging all the balls with the same α value. For the first step, a sphere b_i is removed if:

- The sphere b_i contributes some surface to the model.
- The volume that b_i contributes is less than the feature variable (K) multiply the average contribution of all the spheres.
- The sphere b_i contributes less than the density variable (J) of its volume to the model.
- Topology is preserved.

For the feature variable (K), a value greater than 0.5 and less than 1.5 is recommended. For the density valuable (J), a value between 0 and 0.5 is suggested. In Figure 6 and 7, we show experiments comparing the original GSD versus the SGSD. Similar deformation processes can be obtained with proper choice of variables K and J.

After the deletion, we enlarge the radius of each ball from $\sqrt{w_i}$ to $\sqrt{w_i + \alpha}$ with a value of α that gives the least difference in symmetric difference of the volumes between the unions of the original and simplified set of balls. This increment by α does not change the underlying Voronoi complex. After this, the steps are repeated until the symmetric difference stable.

3.2 Degeneracies

In Deformation I and III, we keep weighted point positions in B_0' the same as in B_0 and change all the weights by the same value α. The benefits for this are: 1. to avoid skinny intermediate mixed cells, and 2. to reduce the complexity of intermediate complex. For example, in Figure 2, in $\mathbb{R}^2$ we deform a skin curve skin (B_0) with four weighted points to its simplified model skin (B_0') by deleting one weighted point in B_0. The intermediate complex is built base on the *super-imposition* of the two Voronoi complex of B_0 and B_0', namely $V(B(t))$ [6]. The complexity of the intermediate complex is equivalent to the number of simplices in $V(B(t))$. In Figure 2, there are in total four intermediate Voronoi vertices in the intermediate complex. However, if we allow different point positions in B_0' or different α changes, after the super-imposition, there are more than 4 intermediate Voronoi vertices (Figure 3).

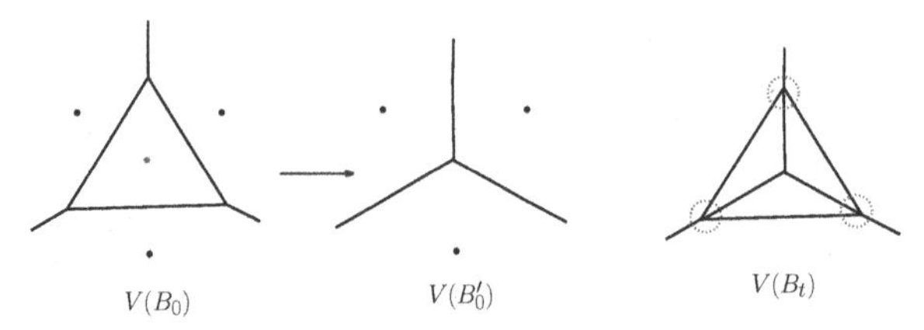

Fig. 2. A four weighted point set B_0 deforms to a subset B_0' by deleting one of its weighted point. We show the Voronoi diagram of B_0, B_0' and the intermediate weighted point set $B(t)$. The dot circled vertices in $V(B(t))$ are intermediate Voronoi vertices with trapezoid intermediate Delaunay cells.

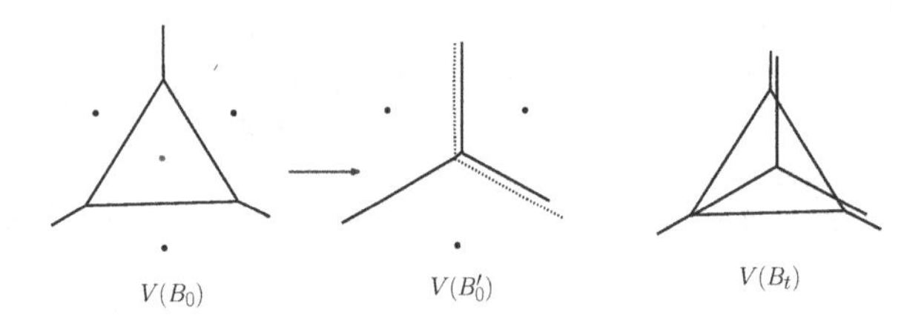

Fig. 3. We give an α to the weighted point on the up-right corner. After super-imposition, we have 6 intermediate Voronoi vertices.

Degeneracies arise in the intermediate complexes in Deformation I and III. The GPA is again unavoidably violated while B_0 and B_0' share the same point locations (The first violation happens in Section 2.2). First, we illustrate the degenerate case in $\mathbb{R}^2$. In $\mathbb{R}^2$, after we super-impose the Voronoi complexes in Deformation I and III, there are Voronoi edges from different Voronoi complex intersect at one of their end-points (See Figure 2 as an example). This degenerate case is represent as tuple (1,0,0). The corresponding intermediate Delaunay cell is a trapezoid which is obtained by deforming a Delaunay edge to a Delaunay triangle (Figure 4).

In $\mathbb{R}^3$, there are two more degenerate cases in the intermediate complex (Table 1). First, Tuple (2,1,1) indicates that a Voronoi face and a Voronoi edge from different Voronoi complexes intersect at one boundary edge of the Voronoi face. For example, in Figure 5(a), we delete one weighted point from a five weighted

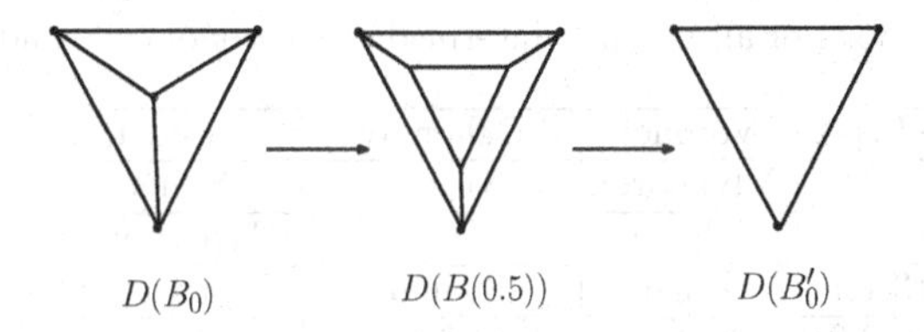

Fig. 4. The intermediate Delaunay triangulation at $t = 0.5$ (middle) with the original simplification shown in Figure 2

point set. There are six such degenerate intermediate Voronoi edges, each of which is shared by four Voronoi regions. The corresponding intermediate Delaunay cells are trapezoids. Second, Tuple (1,0,0) represents an intersection between a Voronoi edge and a Voronoi point from different Voronoi complexes. The corresponding intermediate Delaunay cell is a frustum which deforms from a triangle to a tetrahedron (Figure 5(b)).

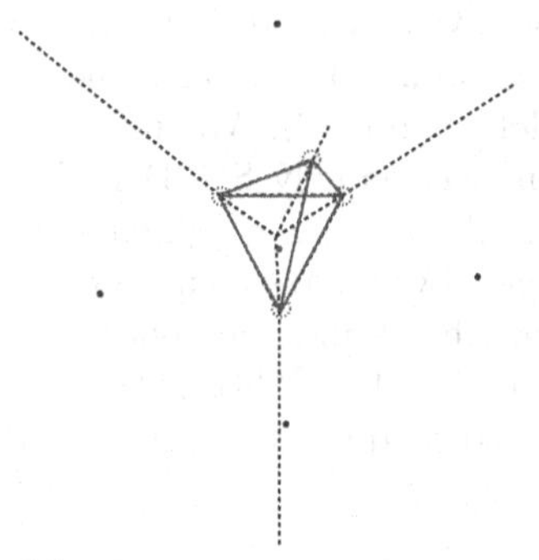

(a) The degenerate Voronoi complex consists of four degenerate Voronoi vertices (circled in red).

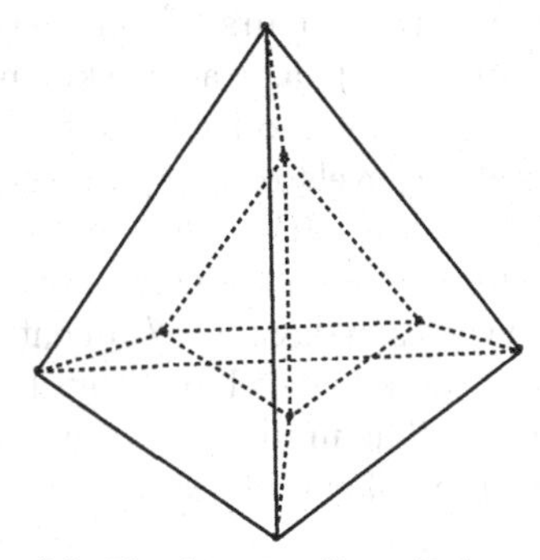

(b) The intermediate Delaunay complex consists of four frustum and one tetrahedron.

Fig. 5. Degenerate intermediate Voronoi and Delaunay complexes after deleting one point in a five weighted point set

3.3 Complexity Analysis

We argue that the time complexity for the new SGSD is $O(m^2+n^2)$ by assuming m' is a user input constant and it is comparatively small to m. The number of weighted points in Deformation I is mm' and we can treat it as $O(m)$ while m' is a small constant. Treating Deformation III similarly, the overall complexity is $O(m^2) + O((m'n')^2) + O(n^2) = O(m^2 + n^2)$.

Table 1. Combinations of all possible intermediate Voronoi cells and Delaunay Cells

Type	Voronoi	Delaunay	Patch
(3,3,3)	Polyhedron	Vertex	Sphere
(3,2,2)	Polygon	Edge	Hyperboloid
(3,1,1)	Edge	Triangle	Hyperboloid
(3,0,0)	Vertex	Tetrahedron	Sphere
(2,2,1)	Edge	Parallelogram	Hyperboloid
(2,1,0)	Vertex	Triangle Prism	Sphere
(2,1,1)	**Edge**	**Trapezoid**	**Hyperboloid**
(1,0,0)	**Vertex**	**Tri. Frustum**	**Sphere**

4 Experiment Result

Experimental results show that SGSD algorithm largely reduces the number of intermediate mixed cells, the number of schedules and the overall running time over the GSD algorithm. We perform the deformations and compare the results over three pairs of skin models: a bunny skin model and a cow skin model (Figure 6), a dragon skin model to a bunny skin model and a mannequin head skin model to a fist skin model (Figure 7). We introduce two simplification levels to compare their performance, namely SGSD-1 ($K = 0.6$, $J = 0.1$) and SGSD-2 ($K = 1.0$, $J = 0.3$). All starting weighted point sets are get from the power crust project developed by Nina Amenta et al. [2] at http://www.cs.ucdavis.edu/ amenta/powercrust.html and the sphere-tree construction toolkit developed by Bradshaw et al. [4] at http://isg.cs.tcd.ie/ spheretree/. The input skin meshes are generated by quality skin mesh software developed by Cheng and Shi [8].

Table 2. Number of intermediate mixed cells for different deformation models

Model pair	GSD	SGSD-1	SGSD-2
Mannequin $\leftrightarrow$ Fist	149,381,672	913,053	317,192
Bunny $\leftrightarrow$ Cow	577,614	182,254	94,911
Dragon $\leftrightarrow$ Bunny	891,161	221,658	103,746

We test both GSD and SGSD algorithms in a 32-bit windows machine with Intel Duo Core 2.33GHz and 4GB RAM. First, a comparison of the total number of intermediate mixed cells is made, as shown in Table 2. The total number of intermediate mixed cells in all three sub-deformations in SGSD are also shown in Table 2. The result shows that the intermediate complex is significantly simplified in SGSD. Second, the whole deformation process is divided into $1,000$ frames and the average number of schedules (Table 3) and average running time (Table 4) are compared in each frame for both algorithms. Based on the statistics collected, both the number of schedules and running time are reduced due to the simpler intermediate complexes in SGSD.

Table 3. Average number of schedules

Model pair	GSD	SGSD-1	SGSD-2
Mannequin ↔ Fist	-	16,491	5,719
Bunny ↔ Cow	10,855	3,712	1,731
Dragon ↔ Bunny	13,774	4,073	2,442

The number of intermediate weighted points is maintained at less than 3 million in GSD. When the number of intermediate weighted points exceeds 3 million, for example, the direct GSD between the mannequin and fist skin models in Figure 1, it is impossible for a 32-bit machine to handle such a large intermediate complex (indicated by the empty cell in Table 2 and 3). However, It is possible to deform the mannequin model to the fist model using the SGSD algorithm by two simplification deformations and one GSD (Figure 7).

The results of SGSD are very similar to the original GSD algorithm as the simplification processes guarantees the volume difference to be small (Figure 6 and 7). Although the simplified objects may lose sharp features of the original objects, it is visually tolerable since it is only one intermediate frame in the whole deformation process.

Table 4. Average time taken in each frame

Model pair	GSD	SGSD-1	SGSD-2
Mannequin ↔ Fist	-	3.18 sec	1.38 sec
Bunny ↔ Cow	5.40 sec	1.87 sec	0.65 sec
Dragon ↔ Bunny	6.14 sec	1.92 sec	0.75 sec

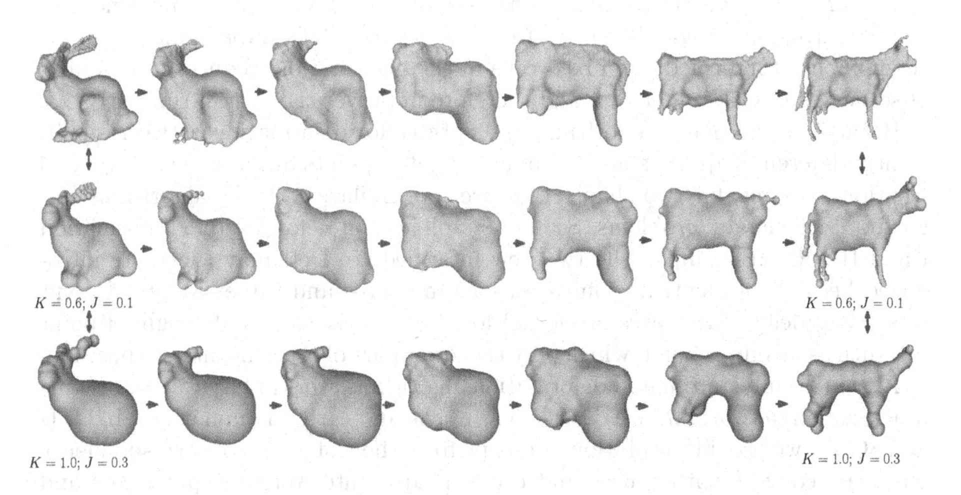

Fig. 6. Different simplification level break down for deformation between bunny and cow

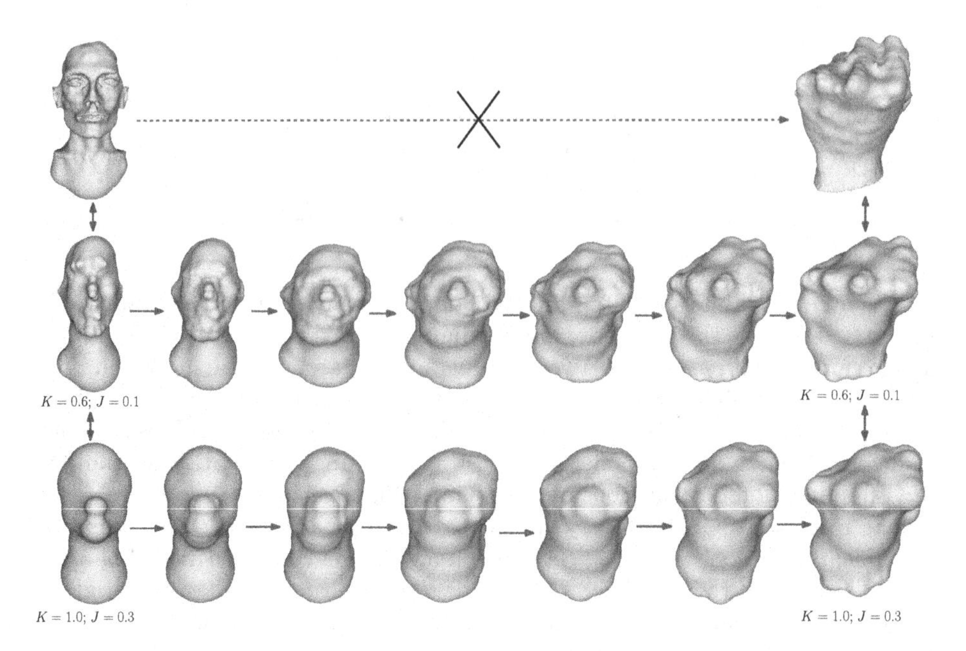

Fig. 7. Different simplification level break down for deformation between mannequin head and fist

5 Conclusion and Discussion

In this paper, we introduce a new simplified general skin deformation algorithm to improve the efficiency of the original GSD algorithm and provide a solution to deform skin surface with huge weighted point sets which originally is impossible to be handled by GSD. We manage to improve the complexity of intermediate complex from the original $O(m^2n^2)$ to $O(m^2 + n^2)$. This improvement reduces the number of intermediate mixed cells in the deformation, reduces the number of schedules in GSD algorithm and therefore improves the overall running time.

Both GSD and SGSD algorithms are suitable for deformation between significantly different objects, since all input weighted points are assumed in general positions. In fact, for two shapes that are too similar, both the algorithms suffer from degeneracy problems. See the simplification deformation (Deformation I and III) as an example, the two new intermediate Delaunay types are introduced because of identical point positions in source and target weighted point sets. More degenerate cases are found for partial movements of weighted point set, such as an elbow bend, which only the lower part of the arm changes position.

Although fully automated deformation algorithms are still far away to become practical in real world industries, such as movie and cartoon animations. In GSD and SGSD, we provide a solution to escape from the restriction of correspondence mapping, convert both source and target shapes into weighted point sets and interpolate the weighted points from the two sets. The limitation of this approach is that we may lose sharp features in the intermediate shapes. This limitation induces our future work, which is to introduce additional reference shapes during the deformation [6].

References

1. Alexa, M., Sorkine, O.: As-rigid-as-possible shape interpolation. In: Proc. SGP (2007)
2. Amenta, N., Choi, S., Kolluri, R.: The power crust. In: Proceedings of 6th ACM Symposium on Solid Modeling, pp. 249–260 (2001)
3. Baran, I., Popovic, J.: Automatic rigging and animation of 3d characters. ACM TOG 26(3) (2007)
4. Bradshaw, G., O'Sullivan, C.: Sphere-tree constuction using medial axis approximation. In: SIGGRAPH Symposium on Computer Animation SCA (2002)
5. CGAL, `http://www.cgal.org`
6. Chen, C., Cheng, H.-L.: Superimposing voronoi complexes for shape deformation. Int. J. Comput. Geometry Appl. (2006)
7. Cheng, H.-L., Dey, T.K., Edelsbrunner, H., Sullivan, J.: Dynamic skin triangulation. Discrete Comput. Geom. (2001)
8. Cheng, H.-L., Shi, X.: Quality tetrahedral mesh generation for macromolecules. In: Asano, T. (ed.) ISAAC 2006. LNCS, vol. 4288, pp. 203–212. Springer, Heidelberg (2006)
9. Cheng, H.-L., Yan, K.: Mesh deformation of dynamic smooth manifolds with surface correspondences. In: Hliněný, P., Kučera, A. (eds.) MFCS 2010. LNCS, vol. 6281, pp. 677–688. Springer, Heidelberg (2010)
10. Edelsbrunner, H.: Deformable smooth surface design. Discrete Comput. Geom., 87–115 (1999)
11. Edelsbrunner, H., Mücke, E.P̃.: Three-dimensional alpha shapes. ACM Trans. Graphics, 43–72 (1994)
12. Kircher, S., Garland, M.: Free-form motion processing. ACM TOG 27(2) (2008)
13. Taylor, R.D., Jewsbury, P.J., Essex, J.W.: A review of protein-small molecule docking methods. J. Comput. Aided Mol. Des.
14. Wang, R., Zhou, K., Snyder, J., Liu, X., Bao, H., Peng, Q., Guo, B.: Variational sphere set approximation for solid objects. Visual. Comput. (2006)
15. Weber, O., Sorkine, O., Lipman, Y., Gotsman, C.: Contex-aware skeletal shape deformation. Computer Graphics Forum 26(3) (2007)
16. Zhang, H., Sheffer, A., Cohen-Or, D., Zhou, Q., van Kaick, O., Tagliascchi, A.: Deformation-driven shape correspondence. In: Proc. SGP (2008)

Author Index